BTEC SERIES

ELECTRICAL AND ELECTRONIC PRINCIPLES II

A. NICOLAIDES, B.Sc., (Eng.), C.Eng.M.I.E.E.
Senior Lecturer and Course Tutor of
THE BTEC NATIONAL DIPLOMA IN
ENGINEERING

*L*ewisham College
previously known as
South East London College

P.A.S.S. PUBLICATIONS

Private Academic & Scientific Studies Ltd.

621.3

PRIVATE ACADEMIC & SCIENTIFIC STUDIES LTD

11 Baring Road, London, SE12 OJP

OTHER TITLES BY THE SAME AUTHOR

BTEC   MATHEMATICS II

GCE 'A'   PURE MATHEMATICS   COMPLEX NUMBERS

GCE 'A'   PURE MATHEMATICS   TRIGONOMETRY

Dedicated to my wife
ALIKI
to whom I am grateful

PREFACE

This book, which is part of the BTEC series in MATHEMATICS AND ELECTRONICS for technicians covers thoroughly the topic of ELECTRICAL AND ELECTRONIC PRINCIPLES II (unit number U86 /329).

The ELECTRICAL AND ELECTRONIC PRINCIPLES II book, like all the books in the series, is divided into two parts. In PART I the theory of ELECTRICAL AND ELECTRONIC PRINCIPLES II which consists of eight chapters is extensively dealt with together with 150 worked examples and exercises fully worked out. A step by step approach is adopted in all the worked examples. There are over two hundred and sixty well illustrated diagrams, a special feature of this book.

PART II of the book acts as a problem solver for all the exercises set at the end of each chapter in PART I.

I am very grateful to my colleague and senior lecturer, Mr. Eddie Coode for checking the manuscript and making useful suggestions.

Anthony Nicolaides

CONTENTS

ELECTRICAL AND ELECTRONIC PRINCIPLES II
UNIT NUMBER U86/329

BTEC

PART I

PART II

THEORY PART I

ELECTRICAL AND ELECTRONIC PRINCIPLES II
UNIT NUMBER U86/329
BTEC

CIRCUIT THEORY

1. *Applies circuit theory to the solution of simple circuit problems.*
    a. *Applies ohm's law to the solution of problems relating to series-parallel combinations of resistors.*

THE UNIT OF CURRENT

The unit of current is the ampere, abbreviated by the symbol A. The ampere may be defined in terms of the force between conductors.

An ampere is defined as the constant current that flows in each of two infinitely-long parallel straight conductors of negligible cross-sectional area separated by a distance of one metre in *vacuo* when the force between these conductors is equal to 2×10^{-7} newtons per metre length. This definition is simply illustrated in Fig. 1.

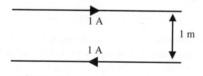

Fig. 1   Two infinitely long conductors placed 1 m apart.

If the currents flow in the same direction, the force is attractive, if the currents flow in the opposite direction as shown in Fig. 1, the force is repulsive.

CURRENT FLOWS DUE TO THE EXISTENCE OF A POTENTIAL DIFFERENCE (VOLTAGE) BETWEEN TWO POINTS IN AN ELECTRICAL CONDUCTOR.

Electrical current is the flow of electrical charge and current flows through a resistor when a source of energy is supplied.

Let us consider a circuit where a d.c. source of e.m.f. E and negligible internal resistance is connected across a resistor R, as shown in Fig. 2.

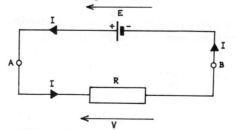

Fig. 2   Conventional current and voltage.

Conventionally, it is assumed that the current *I* flows from the positive terminal of the e.m.f. source to the negative terminal as shown in Fig. 2.

It is observed that the current *I* flows from A to B via the resistor, therefore the point A is positve with respect to the point B, and the potential difference across *R* is deuoted by *V*.

The arrowhead denotes that A is positive and the tail end of the arrow denotes that B is negative. Fig. 3 shows the convention of current and potential difference in a passive component, the resistor.

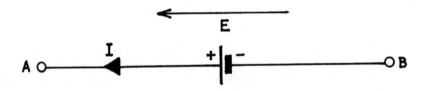

Fig. 3   Conventional current and voltage in a passive component.

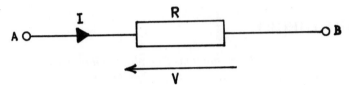

Fig. 4   Conventional current and voltage in an active component.

The cell has an e.m.f. (electromotive force) *E* as shown in Fig. 4, the positive side is denoted by the arrowhead and the negative side is denoted by the tail end of the arrow. Observe that in Fig. 3, *I* and *V* are opposing each other and in Fig. 4, *I* and *E* are in the same direction. Note that current flows only if the circuit is closed. Refering to Fig. 2, we have $E - V = 0$ or $E - IR = 0$ or $E = V = IR$.

THE POTENTIAL DIFFERENCE (P.D.)

The potential difference between two points in a circuit is the electrical pressure or voltage required to cause the current flow between the points. The unit of potential difference in the volt, *V*. The *volt* is defined as the p.d. across a resistance of one ohm carrying a current of one ampere. The p.d. between two points is one volt if one joule of energy is used in sending one coulomb of charge between them.

$$1 \text{ volt} = \frac{1 \text{ joule}}{1 \text{ coulomb}}.$$

The unit of quantity of electric charge is the coulomb. One coulomb is the quantity of electricity conveyed by a current of one empere flowing for one second.

OHM'S LAW

Ohm's Law states that the potential difference across a conductor is proporional to the current flowing through the conductor, provided the tempereature is constant, that is

$V \propto I$

where $\propto$ is the proportionality sign and this sign is replaced by an equal sign by introducing a constant, R, in the equation

$V = RI$

where R = the resistance of the conductor.

$$R = \frac{V}{I} = \frac{\text{Potential difference accros the conductor}}{\text{Current through the conductor}}.$$

LINEAR AND NON LINEAR RESISTORS

For linear resistors, the ratio of V over I (V/I) is always the same and it is equal to the value of the linear resistance, R.

For non-linear resistors, the ratio of V/I is different at different values of V or at different values of I. The following worked examples are used to illustrate the difference between linear and non linear resistors.

WORKED EXAMPLE 1

Tests on linear and non-linear resistors resulted in the following readings:-

V (volts)	0	15	25	35	45	60	
I (mA)	0	7.5	12.5	17.5	22.5	30	} LINEAR
R (Ω)	( )	( )	( )	( )	( )	( )	
I (mA)	0	2	7.5	22.5	44	80	} NON LINEAR
R (Ω)	( )	( )	( )	( )	( )	( )	

Complete the table and plot the following graphs:
(i) I against V for the linear and non linear resistors.
(ii) R against V for the linear and non linear resistors.
Using the above graphs draw your conclusions.

SOLUTION 1

(i) LINEAR RESISTOR

$$R = \frac{V}{I} = \frac{15}{7.5 \times 10^{-3}} = 2,000 \ \Omega \text{ at } V = 15 \text{ volts}$$

$$R = \frac{V}{I} = \frac{25}{12.5 \times 10^{-3}} = 2,000 \ \Omega \text{ at } V = 25 \text{ volts}$$

$$R = \frac{V}{I} = \frac{35}{17.5 \times 10^{-3}} = 2{,}000 \ \Omega \text{ at } V = 35 \text{ volts}$$

$$R = \frac{V}{I} = \frac{45}{22.5 \times 10^{-3}} = 2{,}000 \ \Omega \text{ at } V = 45 \text{ volts}$$

$$R = \frac{V}{I} = \frac{60}{30 \times 10^{-3}} = 2{,}000 \ \Omega \text{ at } V = 60 \text{ volts.}$$

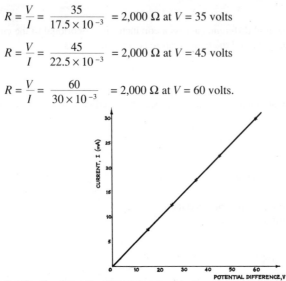

Fig. 5   Graph of I against V, the gradient is the same everywhere and the relationship is linear and passes through the origin.

A graph of I in mA is plotted against V (volts). It is linear and passes through the origin. The gradient is the same everywhere and the ratio V/I is 2,000 Ω. Even when the current through R is zero and V is zero, the resistance is 2,000 Ω.

NON LINEAR RESISTOR

$$\text{At } V = 15 \text{ volts, } R = \frac{V}{I} = \frac{15}{2 \times 10^{-3}} = 7{,}500 \ \Omega$$

$$\text{at } V = 25 \text{ volts, } R = \frac{V}{I} = \frac{25}{7.5 \times 10^{-3}} = 3{,}333 \ \Omega$$

$$\text{at } V = 35 \text{ volts, } R = \frac{V}{I} = \frac{35}{22.5 \times 10^{-3}} = 1{,}555 \ \Omega$$

$$\text{at } V = 45 \text{ volts, } R = \frac{V}{I} = \frac{45}{44 \times 10^{-3}} = 1{,}024 \ \Omega$$

$$\text{at } V = 60 \text{ volts, } R = \frac{V}{I} = \frac{60}{80 \times 10^{-3}} = 750 \ \Omega.$$

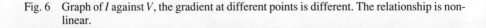

Fig. 6   Graph of I against V, the gradient at different points is different. The relationship is non-linear.

The graph of *I* (mA) against *V* (volts) is non linear and the gradient at the five different values of *V* are not the same.

(i) FOR THE LINEAR CASE

  R (Ω) is plotted against *V* (volts) and is shown to be horizontal line, indicating that the resistance is constant.

(ii) FOR THE NON LINEAR CASE

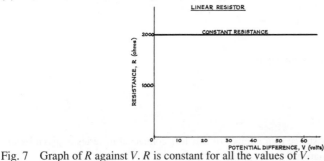

Fig. 7   Graph of *R* against *V*. *R* is constant for all the values of *V*.

  R (Ω) is plotted against *V* (volts) and is shown to be a curve, indicating that the resistance decreases as the voltage is increasing.

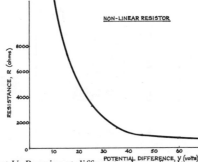

Fig. 8   Graph of *R* against *V*, *R* varies at different values of voltage.

SERIES RESISTORS

Three resistors in series can be connected as shown in Fig. 9, Fig. 10, Fig. 11 and Fig. 12

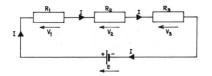

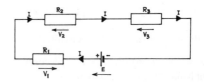

Fig. 9   Resistors in series.

Fig. 10   Resistors in series.

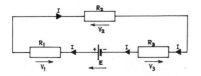

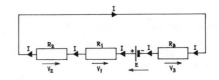

Fig. 11   Resistors in series.

Fig. 12   Resistors in series.

In all the circuits in Figs. 9, 10, 11 and 12 above, it is observed that the current through each resistor is the same, therefore the resistors are in series. The sum of the p.d.s around a closed loop is zero or the sum of the voltages in a series circuit is equal to the total applied e.m.f.

$$V - IR_1 - IR_2 - IR_3 = 0$$

$$\text{or } V = IR_1 + IR_2 + IR_3$$

taking I as a common factor

$$V = I(R_1 + R_2 + R_3).$$

Let R_T be the total resistance of the circuit

$$V = IR_T$$

therefore $R_T = R_1 + R_2 + R_3$.

The total or equivalent resistance of a number of resistors connected in series is given by the sum of the individual resistances.

RESISTORS IN PARALLEL

Three resistors in parallel can be connected as shown in Fig. 13

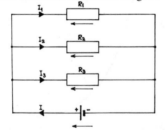

It is observed that the potential difference across each resistor is V, therefore the three resistors are in parallel.

The sum of the currents in resistors connected in parallel is equal to the current into the parallel network

$$I = I_1 + I_2 + I_3.$$

If R_T is the total resistance of the circuit, then $I = V/R_T$ and hence $\dfrac{V}{R_T} = \dfrac{V}{R_1} + \dfrac{V}{R_2} + \dfrac{V}{R_3}$

where $I_1 = V/R_1$, $I_2 = V/R_2$, $I_3 = V/R_3$, according to ohm's law. Taking V as a common factor on the right hand side of the equation

$$\frac{V}{R_T} = V\left(\frac{1}{R_1} + \frac{1}{R_2} + \frac{1}{R_3}\right)$$

dividing both sides of the equation by V

$$\frac{1}{R_T} = \frac{1}{R_1} + \frac{1}{R_2} + \frac{1}{R_3} \qquad \dots (1)$$

The equivalent resistance R_T, of three resistors connected in parallel is given by the equation above.

If the number of resistors is extended to n then $\dfrac{1}{R_T} = \dfrac{1}{R_1} + \dfrac{1}{R_2} + \dfrac{1}{R_3} + \ldots + \dfrac{1}{R_n}$.

For $n = 2$

$$\frac{1}{R_T} = \frac{1}{R_1} + \frac{1}{R_2} \cdot$$

It is convenient to re-arrange this formula as $\dfrac{1}{R_T} = \dfrac{R_2 + R_1}{R_1 R_2}$ or $R_T = \dfrac{R_1 R_2}{R_1 + R_2} \cdot$

Thus the total resistance of two resistors in parallel is expressed as the product of the two resistors over their sum. This formula is *only* applicable for two resistors in parallel. This formula is useful to remember as

$$R_T = \frac{\text{product}}{\text{sum}} \cdot$$

Re-arranging equation (1)

$$\frac{1}{R_T} = \frac{1}{R_1} + \frac{1}{R_2} + \frac{1}{R_3} = \frac{R_2 R_3 + R_1 R_3 + R_1 R_2}{R_1 R_2 R_3}$$

taking the reciprocal of each side, we have

$$R_T = \frac{R_1 R_2 R_3}{R_2 R_3 + R_1 R_3 + R_1 R_2} \cdot$$

This formula is not so easy to remember, but useful for 3 resistors in parallel.

Taking the reciprocal of each side of equation (1)

$$R_T = \frac{1}{\dfrac{1}{R_1} + \dfrac{1}{R_2} + \dfrac{1}{R_3}} \cdot$$

SERIES AND PARALLEL RESISTORS NETWORKS

WORKED EXAMPLE 2

Determine the total resistance of the networks between A and B showing clearly all the steps of your solution and expressing the results in three significant figures.

Network of resistors in series/parallel.

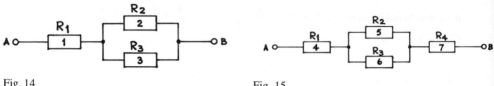

Fig. 14

Fig. 15

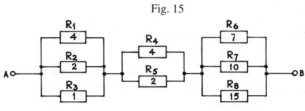

Fig. 16

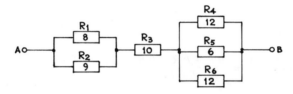

Fig. 17

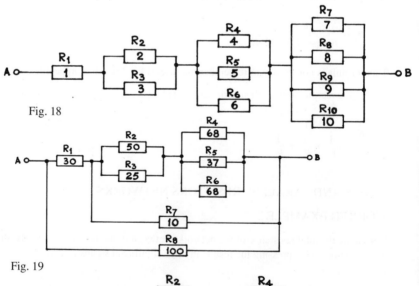

Fig. 18

Fig. 19

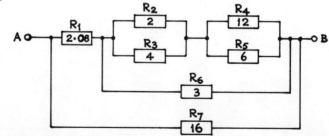

Fig. 20

SOLUTION 2

(i) $R_{AB} = R_1 + \dfrac{R_2 R_3}{R_2 + R_3} = 1 + \dfrac{2 \times 3}{2 + 3} = 1 + \dfrac{6}{5} = 2.2 \, \Omega$

(ii) $R_{AB} = R_1 + \dfrac{R_2 R_3}{R_2 + R_3} + R_4 = 4 + \dfrac{5 \times 6}{5 + 6} + 7 = 1 + 2.73 + 7 = 13.7 \, \Omega$

(iii) $R_{AB} = \dfrac{R_1 R_2}{R_1 + R_2} + R_3 + \dfrac{R_4 R_5 R_6}{R_4 R_5 + R_4 R_6 + R_5 R_6} = \dfrac{8 \times 9}{8 + 9} + 10 + \dfrac{12 \times 6 \times 12}{72 + 72 + 144}$

$R_{AB} = 4.24 + 10 + 3 = 17.2 \, \Omega.$

An easier way to find the combination of the three resistors in parallel is to use the fact that, the total resistance of two equal resistors in parallel is equal to the half value of one, thus 12 Ω in parallel with 12 Ω is a 6 Ω resistor then 6 Ω in parallel with the third 6 Ω resistor gives the combination 3 Ω.

(iv) $R_{AB} = \dfrac{R_1 R_2 R_3}{R_1 R_2 + R_1 R_2 + R_2 R_3} + \dfrac{R_4 R_5}{R_4 + R_5} + \dfrac{R_6 R_7 R_8}{R_6 R_7 + R_7 R_8 + R_6 R_8}$

$R_{AB} = \dfrac{4 \times 2 \times 1}{4 \times 2 + 4 \times 1 + 2 \times 1} + \dfrac{4 \times 2}{4 + 2} + \dfrac{7 \times 10 \times 15}{7 \times 10 + 7 \times 15 + 10 \times 15}$

$R_{AB} = 0.571 + 1.33 + 3.23 = 5.13 \, \Omega$

$R_{AB} = 5.13 \, \Omega.$

(v) $R_{AB} = R_1 + \dfrac{R_2 R_3}{R_2 + R_3} + \dfrac{R_4 R_5 R_6}{R_4 R_5 + R_4 R_6 + R_5 R_6} + \dfrac{R_7 R_8 R_9 R_{10}}{R_7 R_8 R_9 + R_7 R_8 R_{10} + R_7 R_9 R_{10} + R_8 R_9 R_{10}}$

$R_{AB} = 1 + \dfrac{2 \times 3}{2 + 3} + \dfrac{4 \times 5 \times 6}{4 \times 5 + 4 \times 6 + 5 \times 6} + \dfrac{7 \times 8 \times 9 \times 10}{7 \times 8 \times 9 + 7 \times 8 \times 10 + 7 \times 9 \times 10 + 8 \times 9 \times 10}$

$R_{AB} = 1 + 1.2 + 1.62 + 2.08 = 5.91 \, \Omega.$

The four parallel resistors can be found easier as follows:-

$\dfrac{1}{R_T} = \dfrac{1}{R_7} + \dfrac{1}{R_8} + \dfrac{1}{R_9} + \dfrac{1}{R_{10}} = \dfrac{1}{7} + \dfrac{1}{8} + \dfrac{1}{9} + \dfrac{1}{10} = 0.143 + 0.125 + 0.111 + 0.1 = 0.479$

taking the reciprocals, we have

$R_T = 2.08 \, \Omega.$

(vi) $R_{T_1} = \dfrac{R_2 R_3}{R_2 + R_3}$ $\qquad\qquad R_{T_2} = \dfrac{R_4 R_5 R_6}{R_4 R_5 + R_4 R_6 + R_5 R_6}$

$R_T = R_{T_1} + R_{T_2}$

$$R_1 + \frac{(R_{T_1} + R_{T_2})\, R_7}{R_{T_1} + R_{T_2} + R_7} \;=\; R = \frac{R_T R_7}{R_T + R_7} + R_1.$$

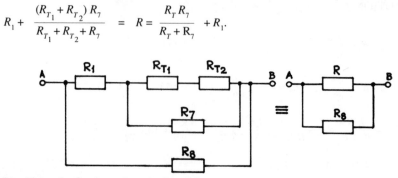

Fig. 21   Network of resistors in series/parallel.

$$R_{T_1} = \frac{R_2 R_3}{R_2 + R_3} = \frac{50 \times 25}{50 + 25} = 16.7\ \Omega$$

$$R_{T_2} = \frac{R_4 R_5 R_6}{R_4 R_5 + R_4 R_6 + R_5 R_6} = \frac{68 \times 37 \times 68}{68 \times 37 + 68 \times 68 + 37 \times 68} = 17.7\ \Omega$$

$$R_T = R_{T_1} + R_{T_2} = 16.7 + 17.7 = 34.4\ \Omega$$

$$R = \frac{R_T R_T}{R_T + R_7} + R_1 = \frac{34.4 \times 10}{34.4 + 10} + 30 = 7.75 + 30 = 37.8\ \Omega$$

$$R = 37.8\ \Omega$$

$$R_{AB} = \frac{R R_8}{R + R_8} = \frac{37.8 \times 100}{37.8 + 100} = 27.4\ \Omega.$$

(vii) $R_{T_1} = \dfrac{R_2 R_3}{R_2 + R_3}$ $\qquad\qquad$ $R_{T_2} = \dfrac{R_4 R_5}{R_4 + R_5}$ $\qquad\qquad$ $R_T = R_{T_1} + R_{T_2}$

$$R = \frac{R_T R_6}{R_7 + R_6} \qquad\qquad R_{AB} = \frac{(R_1 + R)\, R_7}{R_1 + R + R_7}$$

$$R_{T_1} = \frac{R_2 R_3}{R_2 + R_3} = \frac{2 \times 4}{2 + 4} = \frac{4}{3} = 1.33\ \Omega \qquad\qquad R_{T_2} = \frac{R_4 R_5}{R_4 + R_5} = \frac{12 \times 6}{12 + 6} = \frac{72}{18} = 4\ \Omega$$

$$R_T + R_{T_1} + R_{T_2} = 1.33 + 4 = 5.33\ \Omega$$

$$R = \frac{R_T R_6}{R_T + R_6} = \frac{5.33 \times 3}{5.33 + 3} = 1.92\ \Omega \qquad\qquad R_{AB} = \frac{(2.08 + 1.92)\, 16}{2.08 + 1.92 + 16} = 3.2\ \Omega.$$

b. *Derives and applies the concept of proportional voltage and current division in circuit anualysis.*

THE CURRENT DIVIDER

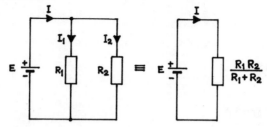

Fig. 22   Current divider. Equivalent circuit.

The total resistance for the circuit is

$$R_T = \frac{R_1 R_2}{R_1 + R_2}$$

$$\frac{V}{I} = \frac{R_1 R_2}{R_1 + R_2}$$

$$V = \frac{R_1 R_2}{R_1 + R_2} I$$

$$I_1 = \frac{V}{R_1} = \frac{R_1 R_2}{R_1 + R_2} I \frac{1}{R_1} = \frac{R_2}{R_1 + R_2} I$$

$$\boxed{I_1 = \frac{R_2}{R_1 + R_2} I} \qquad \ldots (1)$$

$$I_2 = \frac{V}{R_2} = \frac{R_1 R_2}{R_1 + R_2} I \frac{1}{R_2} = \frac{R_1}{R_1 + R_2} I$$

$$\boxed{I_2 = \frac{R_1}{R_1 + R_2} I} \qquad \ldots (2)$$

WORKED EXAMPLE 3

Calculate the currents in the 1 Ω and 2 Ω resistors.

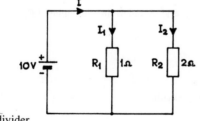

Fig. 23   Current divider.

SOLUTION 3

The total resistance of the circuit

$$R_T = \frac{R_1 R_2}{R_1 + R_2} = \frac{1 \times 2}{1 + 2} = \frac{2}{3} \, \Omega.$$

The total current $I = \dfrac{V}{R_T} = \dfrac{10}{\dfrac{2}{3}} = \dfrac{30}{2} = 15$ A.

Using equations (1) and (2)

we have $I_1 = \dfrac{2}{3}$ (15) = 10 A

$$I_2 = \dfrac{1}{3} \text{ (15)} = 5 \text{ A.}$$

WORKED EXAMPLE 4

The total current in fig 22, $I = 2$ A. If $R_1 = 10$ KΩ and $R_2 = 30$ KΩ, determine the currents I_1 and I_2.

SOLUTION 4

$$I_1 = \dfrac{R_2}{R_1 + R_2} \; I = \dfrac{30 \times 10^3}{(10 + 30) \times 10^3} \times 2 = \dfrac{60}{40} = 1.5 \text{ A}$$

$I_2 = 2 - 1.5 = 0.5$ A.

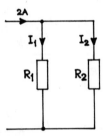

Fig. 24   Current divider.

WORKED EXAMPLE 5

In fig. 22, if $I_1 = 5$ A, $R_2 = 10$ Ω and $I = 12$ A, determine the value of R_1.

SOLUTION 5

$I_2 = I - I_1 = 12 - 5 = 7$ A

p.d. across $R_2 = I_2 R_2 = 7 \times 10 = 70$ volts

p.d. across $R_1 = 70 = I_1 R_1$

$$R_1 = \dfrac{70}{5} \; 14 \text{ Ω}$$

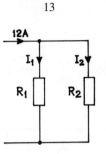

Fig. 25   Current divider.

THE POTENTIAL DIVIDER

The potential divider or potentiometer is used in order to obtain a variable voltage from a constant voltage supply.

Consider the circuit in Fig. 26. The input voltage is fixed (V) and the output voltage (V_1) is taken at the terminals A and B.

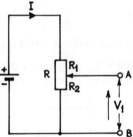

Fig. 26   Potential divider.

By changing the position of the sliding contact A, we can obtain any voltage from zero to the value of V.

$V = I(R_1 + R_2)$ where the sliding contact divides the resistance R into resistances R_1 and R_2

$V_1 = IR_2$

$V = I(R_1 + R_2)$ ... (3)

$V_1 = I R_2$ ... (4)

Dividing equation (3) by equation (4)

$$\frac{V}{V_1} = \frac{I(R_1 + R_2)}{I R_2}$$

we have $\dfrac{V}{V_1} = \dfrac{R_1 + R_2}{R_2}$

or
$$\boxed{V_1 = \frac{R_2}{R_1 + R_2} V}$$

WORKED EXAMPLE 6

If $V = 100$ volts $R_2 = 10$ KΩ, $R_1 = 90$ KΩ, calculate V_1, the p.d. across R_2, and V_2 across R_1.

SOLUTION 6

$$V_1 = \frac{10 \times 10^3}{10 \times 10^3 + 90 \times 10^3} \times 100 = \frac{10}{100} \times 100 = 10 \text{ volts}$$

$$V = V_1 + V_2 = 10 + 90 = 100 \ V.$$

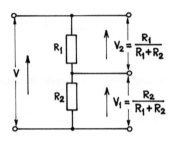

Fig. 27   Potential divider

WORKED EXAMPLE 7

The circuit diagram of a potential divider or 100 Ω is as showm in Fig. 28.

AB is connected across a 100 V d.c. supply and *CD* is connected across a 5 Ω resistor which takes a current of 1 A. Determine the position of the sliding contact, *X*.

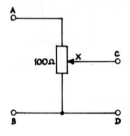

Fig. 28   Potential divider

SOLUTION 7

Let R_1 and R_2 be the resistances of the 100 Ω. The total resistance of the circuit is

$$R_T = R_1 + \frac{R_2 R_L}{R_2 + R_L} \ .$$

The total current, $I = V/R_T$.

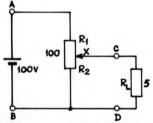

Fig. 29   Loaded potential divider

Let the resistance of $AX = R_1$ then $BX = R_2 = 100 - R_1$

$$\frac{100}{R_1 + \dfrac{(100 - R_1)\,5}{100 - R_1 + 5}} = I$$

the current through $R_L = 1 = \dfrac{R_2}{R_2 + R_L} \; I = \dfrac{100 - R_1}{100 - R_1 + R_L} \; I$

therefore $\dfrac{100 - R_1}{100 - R_1 + 5} \; I = 1$

therefore $\dfrac{100 - R_1}{100 - R_1 + 5} \cdot \dfrac{100}{R_1 + \dfrac{(100 - R_1)\,5}{100 - R_1 + 5}} = 1$

$$\dfrac{(100 - R_1)\,100\,(100 - R_1 + 5)}{(100 - R_1 + 5)\,[R_1(100 - R_1 + 5) + 5\,(100 - R_1)]} = 1$$

$100\,(100 - R_1) = R_1(100 - R_1 + 5) + 5\,(100 - R_1)$

$10{,}000 - 100\,R_1 + 100\,R_1 - R_1^2 + 5\,R_1 + 500 - 5\,R_1$

$R_1^2 - 200\,R_1 + 9{,}500 = 0$

$R_1 = \dfrac{200 \pm \sqrt{200^2 - 4 \times 9{,}500}}{2} = \dfrac{200 \pm 44.72}{2}$

$R_1 = 122.36\ \Omega$ or $R_1 = 77.64\ \Omega$.

The former is disregarded, and therefore $R_1 = 77.64\ \Omega$

WORKED EXAMPLE 8

Calculate R_1 in the network shown in Fig. 30.

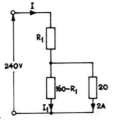

Fig. 30   Loaded potential divider

SOLUTION 8

The p.d. across the 20 Ω resistor $= 20 \times 2 = 40$ V therefore the p.d. across $R_1 = 240 - 40 = 200$ V
the p.d. across the $160 - R_1$ resistor is 40 V the current through the $(160 - R_1)$ resistor is

$I_1 = \dfrac{40}{160 - R_1}$

$I = I_1 + 2 = \dfrac{40}{160 - R_1} + 2 = \dfrac{200}{R_1}$

$\dfrac{40\,R_1}{160 - R_1} + 2\,R_1 = 200$

$$40\,R_1 + 2\,R_1\,(160 - R_1) = 200\,(160 - R_1)$$

$$40\,R_1 + 320\,R_1 - 2\,R_1^2 = 32{,}000 - 200\,R_1$$

$$2\,R_1^2 - 200\,R_1 - 360\,R_1 + 32{,}000 = 0$$

$$R_1^2 - 280\,R_1 + 16{,}000 = 0$$

$$R_1 = \frac{280 \pm \sqrt{280^2 - 4 \times 10{,}000}}{2} = \frac{280 \pm 120}{2}$$

$$R_1 = \frac{500}{2} = 250\ \Omega \qquad\qquad \text{or } R_1 = \frac{160}{2} = 80\ \Omega$$

$R_1 = 80\ \Omega$ is the correct value as $160 - 250$ produces a negative resistance.

SUMMARY 1

OHM'S LAW

THE POTENTIAL DIFFERENCE ACROSS A CONDUCTOR IS PROPORTIONAL TO THE CURRENT FLOWING THROUGH THE CONDUCTOR, PROVIDED THE TEPMERATURE IS CONSTANT.

CURRENT DIVIDER

THE CURRENT THROUGH ONE RESISTOR OF TWO RESISTORS IN PARALLEL IS EQUAL TO THE RATIO OF OTHER RESISTOR OVER THE SUM OF THE TWO RESISTORS TIMES THE TOTAL CURRENT

$$I_1 = \frac{R_2}{R_1 + R_2}\,I$$

$$I_2 = \frac{R_1}{R_1 + R_2}\,I$$

where I_1 is the current through R_1, I_2 is the current through R_2 and I is the total current.

VOLTAGE DIVIDER

THE VOLTAGE ACROSS A RESISTOR OF RESISTANCE R_1 OF TWO RESISTORS IN SERIES IS EQUAL TO THE RATIO OF R_1 TO $R_1 + R_2$ TIMES THE TOTAL VOLTAGE

$$V_1 = \frac{R_1}{R_1 + R_2}\,V$$

$$V_2 = \frac{R_2}{R_1 + R_2}\,V$$

where V_1 is the p.d. across R_1, V_2 is the p.d. across R_2, and V is the total voltage.

EXERCISES 1

1. Three resistors of 1, 2 and 3 Ω are connected in parallel. If the total current taken is 12 A find the current through each resistor.

   (Ans. 6.55 A, 3.28 A, 2.18 A)

2. Two resistors of 3 Ω and 5 Ω are connected in parallel. If the total current taken is 1 A find the current through each resistor.

   (Ans. 0.625 A, 0.375 A)

3. Two resistors of 2 Ω and 8 Ω are connected in parallel. If the total current taken is 5 A find the current through each resistor.

   (Ans. 1 A, 4 A)

4. Determine the currents, I_1, and I_2 shown in Fig. 31.

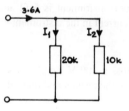

   Fig. 31   Current divider

   (Ans. 1.2 A, 2.4 A)

5. Determine the currents I_1, I_2 and I_3 as shown in Fig. 32.

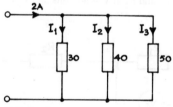

   Fig. 32   Currents in parallel circuit

   (Ans. 0.853 A, 0.64 A, 0.512 A)

6. If the total resistance of the network is 1 Ω, in Fig. 33.

   (i)   Find the value of the resistance, R.

   (ii) Find the currents I_1 and I_2 and the total current taken from the supply, I.

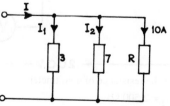

   Fig. 33   Currents in parallel circuit

   (Ans. (i) 1.91 Ω (ii) 6.37 A, 2.37 A, 19.1 A)

7. The total resistance of this combination of resistors is 10 Ω, in Fig. 34. Determine the value of R_1, R_2, R_3 and R_4.

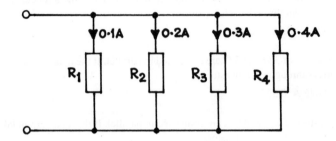

Fig. 34   Resistor in parallel circuit

(Ans. 100 Ω, 50 Ω, 33.3 Ω, 25 Ω)

8. The resistance of a moving coil instrument is 5 Ω and it is provided with a shunt of resistance 0.0005 Ω. Find the current in the instrument when used to measure 15 A.

(Ans. 1.5 mA)

9. A resistor R is connected across a 100 V supply. A voltmeter of resistance 1,500 Ω is connected between the centre of the resistor and one side of the supply. Determine R if the voltmeter reading is 35 V.

(Ans. 2570 Ω)

10. Two resistors are connected in series as shown in Fig. 35 and Fig. 36 and the voltmeters V_1 and V_2 are connected in turn to measure the p.d. across R_1 and across R_2.

Determine the values of R_1 and R_2, if V_1 = 80 volts and V_2 = 100 volts.

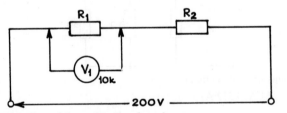
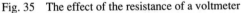

Fig. 35   The effect of the resistance of a voltmeter

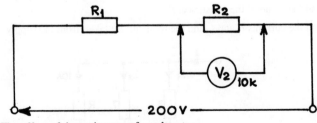

Fig. 36   The effect of the resistance of a voltmeter

(Ans. R_1 = 2,000 Ω, R_2 = 2,500 Ω)

11. Determine the value of the potential difference across the 5 Ω resistor, V_1 in Fig. 37 (Ans. 1.67 V)

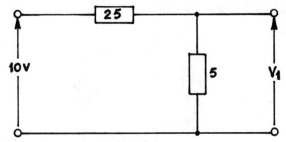

Fig. 37   Potential divider

12. Determine the value of the the the p.d. across the two 10 KΩ resistor, V_1 as shown in Fig. 38.

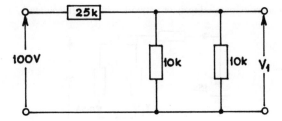

Fig. 38   Loaded potential divider

(Ans. 16.7 V)

13. Determine the value of V by employing
    (i)   the potential divider method
    (ii) the current divider method
    in Fig. 39. (Ans. 4.29 volts)

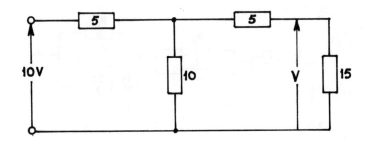

Fig. 39   Ladder network. Potential divider principle.

14. Fig. 40 shows a d.c. source connected across a variable load resistance R. The e.m.f. of the source is 6 V, its internal resistance is 1 Ω. The table below shows the current through the load in amperes, the load resistance in ohms and the power dissipated in the load in watts.

I (A)	6	( )	4	( )	3	( )	2.4	( )	2
R (Ω)	0	$\frac{1}{4}$	$\frac{1}{2}$	$\frac{3}{4}$	1	$1\frac{1}{4}$	$1\frac{1}{2}$	$1\frac{3}{4}$	2
P (W)	0	( )	8	( )	9	( )	8.64	( )	8.

Complete the table and plot P against R and from your graph determine the maximum power dissipated in the load and note the value of R where this power occurs. Draw you conclusions. (Ans. 9 W, $R = r = 1\ \Omega$)

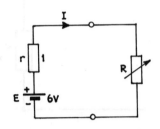

Fig. 40   Maximum power transfer

15. Fig. 41 shows a series-parallel resistor network with the given data.

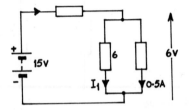

Fig. 41   Loaded potential divider

Calculate the value of R showing clearly the steps of your solution.

(Ans. 6 Ω)

16. (a) Determine the total resistance of the network shown in Fig. 42.

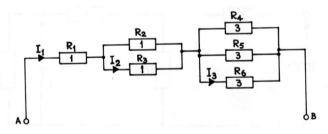

Fig. 42   Network of resistors. Determination of currents.

(b) If the p.d. across AB is 10 volts. Determine: (i) the total current, I_1,

(ii) the current, I_2,

(ii) the current, I_3.

(Ans. (a) 2.5 Ω (b) (i) 4A   (ii) 2 A   (iii) 1.3 A.

KIRCHHOFF'S LAWS

c.   Applies Kirchhoff's Laws to problems involving not more than two unknowns.

NODE OR JUNCTION OF A NETWORK

A node or a junction is shown in Fig. 43 as the bold dot at N. Currents flow into, or out of the node or junction.

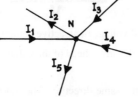

Fig. 43   Many wires connected to a common point (node or junction)

CONNECTION OF CURRENT FLOW

If the currents flowing into the node are assumed to be positive then the currents flowing out of the node are negative. If the currents flowing into the node are assumed to be negative then the currents flowing out of the node are positive.

KIRCHHOFF'S CURRENT LAW (K.C.L.)

The sum of the currents flowing at the junction is equal to zero. Referring to Fig. 43

$$I_1 + I_3 + I_4 - I_2 - I_5 = 0$$

or $$-I_1 - I_3 - I_4 + I_2 + I_5 = 0$$

We can clearly see that it does not matter which currents are positive and which currents are negative. If I_1, I_3, I_4 are positive then I_2, I_5 are negative and if I_1, I_3, I_4 are negative then I_2, I_5 are positive.

LOOP OR MESH OF A NETWORK

A simple loop or mesh is shown in Fig. 44.

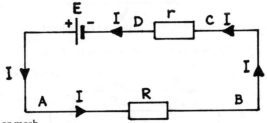

Fig. 44   One loop or mesh

where there is continuity of current and the loop is closed. It starts from the positive pole of the e.m.f., E, through the resistor R, the load, the internal resistance, r, of the cell, and finishes at the negative pole of the e.m.f.

VOLTAGE AND CURRENT CONVENTIONS

The current I is assumed to flow from the positive pole of E to the negative pole of E. The p.ds established in the network of Fig. 44 are one across E, one across R and one across r. Fig. 45

shows the first p.d. across the cell which is an active component.

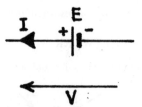

Fig. 45   Conventional current and voltage in an active component.

The positive terminal is shown by the arrow head of the voltage arrow and the negative terminal is shown by the tail end of the arrow.

I and *V* the active device are in the same direction. The other voltages are shown across the passive components, *R* and *r*.

Fig. 46 shows the current flowing through the passive component from A to B, and since the current flows from A to B, A is more positive than B.

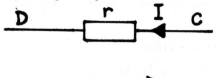

Fig. 46   Conventional current and voltage in a passive component.

I and *V* are in opposition.

Similarly the p.d. across *r* is as shown in Fig. 47. I flows from C to D, and C is more positive than D the arrow head shows the more positive potential.

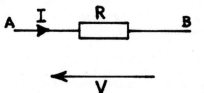

Fig. 47   Conventional current and voltage in a passive component.

The p.d.s around the closed loop of Fig. 48 may be written as $E - IR - Ir = 0$ ... (2)

considering the anticlockwise p.d., *E*, to be positive and clockwise p.d.s *Ir* and *IR* to be negative.

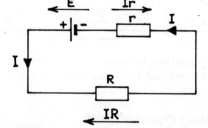

Fig. 48   Conventional currents and voltages in a one loop or mesh circuit.

Alternatively, if we consider the clockwise p.d. positive then the anticlockwise p.d. is negative. This is shown if we change all the signs of equation (2)

$$-E + IR + Ir = 0$$
therefore $$Ir + IR = E$$
or $$I = \frac{E}{r + R}$$

KIRCHHOFF'S VOLTAGE LAW (K.V.L.)

The sum of the e.m.f.s around the closed loop is ZERO.
Referring to Fig. 48.

$$E - IR - Ir = 0$$
or $$E = I\,(R + r).$$

It is very important for the student to practice and understand the conventions of currents and voltages for active and passive components.

For this reason, the student should mark the currents and the p.d.s for various networks shown below.

WORKED EXAMPLE 9

Determine the current in each branch and the magnitude and direction of the p.d. between points A and C, for the circuit in Fig. 49.

Determine also, the values V_{AD} and V_{ED} and hence V_{AE}.

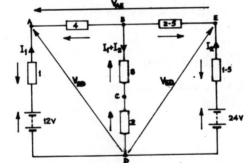

Fig. 49   Conventional currents and voltages in a two loop or mesh circuit.

SOLUTION 9

Assume the currents flow as shown.

Applying *KVL* to loops *ABDA* and *BEDB, ABDA*

$$12 - I_1\,(1) - 4I_1 - 8\,(I_1 + I_2) - 2\,(I_1 + I_2) = 0$$
or $$15I_1 + 10I_2 = 12$$
or $$\boxed{3I_1 + 2I_2 = 2.4} \qquad \dots (1)$$

BEDB loop

$$2\,(I_1 + I_2) + 8\,(I_1 + I_2) + 2.5I_2 + 1.5I_2 = 24$$
or $$10I_1 + 14I_2 = 24$$

or $\boxed{I_1 + 1.4I_2 = 2.4}$ ... (2)

Solving the simultaneous equations (1) and (2)

$$3I_1 + 2I_2 = 2.4$$

$$I_1 + 1.4I_2 = 2.4 \qquad\qquad \times (-3)$$

$$\overline{}$$

$$3I_1 + 2I_2 = 2.4 \qquad\qquad \text{Adding}$$

$$-3I_1 - 4.2I_2 = -7.2$$

$$-2.2I_2 = -4.8$$

$$I_2 = \frac{2.4}{1.1} = 2.18 \text{ A}.$$

From equation (2), $I_1 + 1.4\,(2.18) = 2.4$

$$I_1 = -0.652$$

$$V_{AC} = 4I_1 + 8\,(I_1 + I_2) = 4\,(-0.652) + 8\,(-0.652 + 2.18)$$

$$V_{AC} = -2.608 + 12.224$$

$$\underline{V_{AC} = 9.62 \text{ volts}.}$$

$$V_{AD} = 4I_1 + 10\,(I_1 + I_2) = 14I_1 + 10I_2$$

$$V_{AD} = 14\,(-0.652) + 10 \times 2.18 = -9.128 + 21.8$$

$$\underline{V_{AD} = 12.7 \text{ volts}.}$$

$$V_{ED} = 2.5I_1 + 10\,(I_1 + I_1) = 10I_1 + 12.5I_2$$

$$V_{ED} = 10I_1 + 12.5I_1 = 10\,(-0.652) + 12.5\,(2.18)$$

$$V_{ED} = -6.52 + 27.25$$

$$\underline{V_{ED} = 20.7 \text{ volts}.}$$

$$V_{AE} + V_{ED} = V_{AD}$$

$$V_{AE} = V_{AD} - V_{ED} = 12.7 - 20.7$$

$$V_{AE} = -8 \text{ volts}.$$

therefore E is positive with respect to A.

WORKED EXAMPLE 10

Draw the assumed direction of branch current in the following circuit diagrams and hence mark the direction of potential differences across each component.

Use small arrows for currents and large arrows for voltages.

SOLUTION 10

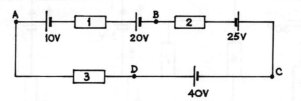

Fig. 50

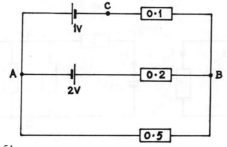

Fig. 51

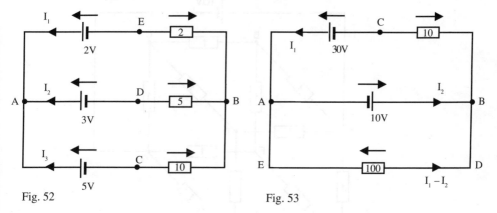

Fig. 52

Fig. 53

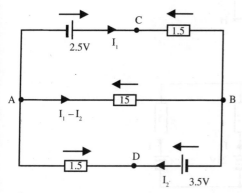

Fig. 54

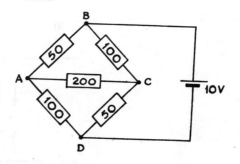

Fig. 55

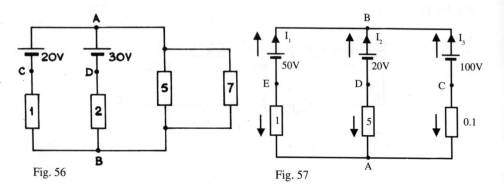

Fig. 56

Fig. 57

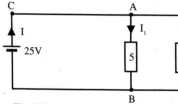

Fig. 58

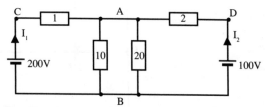

Fig. 59

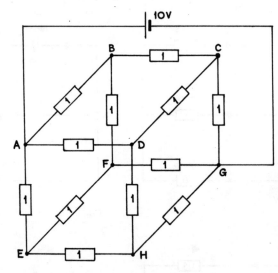

Fig. 60

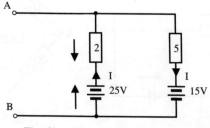

Fig. 61

WORKED EXAMPLE 11

Use Kirchhoff's laws to calculate (a) the three branch currents in the circuit of Fig. 62

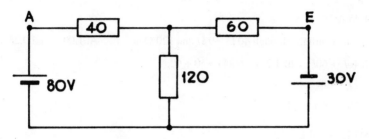

Fig. 62

(b) the potential difference between A and E, stating which point is at the higher potential.

SOLUTION 11

(a) Let the currents be as shown in the diagram of Fig. 63. The p.d. across each component is marked by a large arrow.

Applying Kirchhoff's laws for the loops: ABCDA and BEFCB, we have

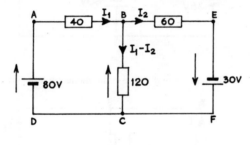

Fig. 63

$ABCDA$ loop

$$80 = 40I_1 + 120\,(I_1 - I_2)$$

or $\quad 160I_1 - 120I_2 = 80$

or $\quad \boxed{4I_1 - 3I_2 = 2} \qquad \ldots (1)$

$BEFCB$ loop

$$120\,(I_1 - I_2) - 60I_2 = -30$$

or $\quad 120I_1 - 180I_2 = -30$

or $\quad 4I_1 - 6I_2 = -1 \qquad \ldots (2)$

Solving the simultaneous equations (1) and (2)

$$4I_1 - 3I_2 = 2 \qquad \ldots (1)$$

$$4I_1 - 6I_2 = -1 \qquad \ldots (2)$$

Equation (1) minus equation (2)

$$3I_2 = 3$$

$$I_2 = 1\ \text{A}.$$

Substituting in equation (1) this value

$$4I_1 - 3(1) = 2$$
$$4I_1 = 5$$
$$I_1 = 1.25 \text{ A}.$$

The three branch currents through 40 Ω, 60 Ω and 120 Ω are: 1.25 A, 1.00 A and 0.25 A.

(b) $V_{AE} = 40I_1 + 60I_2 = 40(1.25) + 60(1) = 50 + 60$

       $V_{AE} = 110$ volts

A is at a higher potential than E.

WORKED EXAMPLE 12

Use Kirchhoff's laws to obtain the values of the currents I_1, I_2 and I_3 in Fig. 64.

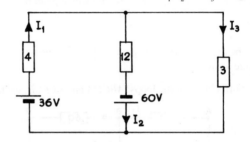

Fig. 64

SOLUTION 12

Mark the voltages across the passive and active components by large arrows, in the Fig. 65.

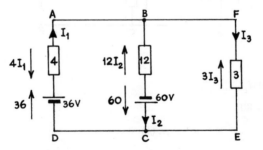

Fig. 65

Loop *ABCDA*

Taking a clockwise direction around this loop

$$36 + 60 - 4I_1 - 12I_2 = 0$$

or $4I_1 + 12I_2 = 96$

or $\boxed{I_1 + 3I_2 = 24}$ ... (1)

Loop **BDECB**

Taking again a clockwise direction around this loop

$$12I_2 - 3I_3 - 60 = 0 \text{ but } I_3 = I_1 - I_2$$

or $12I_2 - 3I_1 + 3I_2 = 60$

or $-3I_1 + 15I_2 = 60$

or $-I_1 + 5I_2 = 20$ ... (2)

Solving the simultaneous equations

$I_1 + 3I_2 = 24$ ... (1)

$-I_1 - 5I_2 = 20$ ... (2)

adding (1) and (2) we have

$8I_2 = 44$

$I_2 = \dfrac{22}{4} = 5.5$ A.

From (1) $I_1 + 3(5.5) = 24$

$\qquad I_1 = 24 - 16.5 = 7.5$ A or $I_2 = 7.5$ A

$\qquad \underline{I_1 = 7.5 \text{ A}}$

$\qquad I_3 = I_1 - I_2 = 7.5 - 5.5 = 2.0$ A

$\qquad \underline{I_3 = 2.0 \text{ A.}}$

WORKED EXAMPLE 13

(a) (i) State the number of junctions or nodes that exist in the circuit of Fig. 66.

(ii) Assuming that the branch currents are flowing as shown in the circuit of Fig. 66, apply Kirchhoff's current law and write down an equation at one of the junctions.

(iii) State and label the number of loops that exist in the circuit of Fig. 66. Apply Kirchhoff's voltage law, and write down two equations of two such loops, in terms of I_1 and I_2.

(iv) Solve the two simultaneous equations you have obtained in (ii) and (iii) giving your answers in three significant figures.

(v) Redraw the circuit of Fig. 66 and indicate the branch currents and p.d.s by small and large arrows.

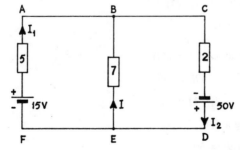

Fig. 66

(b) For the circuit shown in Fig. 67 calculate the resistance between the terminals A and B.

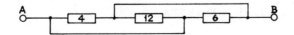

Fig. 67

SOLUTION 13

(a) (i) There are two junctions or nodes, namely the points B and E

(ii) At B

$$I_1 + I - I_2 = 0$$

At E

$$I = I_2 - I_1$$

(iii) Three loops

ABEFA, BCDEB, and ABCDEFA.

Two such loops are:

ABEFA loop

$$15 + 7I = 5I_1$$

$$15 + 7(I_2 - I_1) - 5I_1 = 0$$

$$\boxed{-12I_1 + 7I_2 = -15}$$

BCDEB loop

$$50 = 2I_2 + 7I$$

$$7(I_2 - I_1) + 2I_2 = 50$$

$$\boxed{-7I_1 + 9I_2 = 50}$$

(iv) Solving the equations (1) and (2) simultaneously

$$-7I_1 + 9I_2 = 50 \quad \ldots (1) \times (12)$$

$$-12I_1 + 7I_2 = -15 \quad \ldots (2) \times (-7)$$

Multiplying (1) by 12 and (2) by (–7) we have

$$-84I_1 + 108I_2 = 600 \ldots (3)$$

$$84I_1 - 49I_2 = 105 \quad \ldots (4)$$

Adding equations (3) and (4)

$$59I_2 = 705$$

$$I_2 = 11.95 \text{ A}$$

substituting in (1)

$$-7I_1 + 9(11.95) = 50$$

$$-7I_1 = 50 - 107.55$$

$$I_1 = 8.22 \text{ A}$$

$$I = I_2 - I_1 = 11.95 - 8.22 = 3.73 \text{ A}$$

$$I = 3.73 \text{ A}.$$

(v) The direction of currents are correct as shows in Fig. 68.

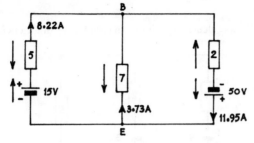

Fig. 68

(b) Mark the terminals as shown in Fig. 69.

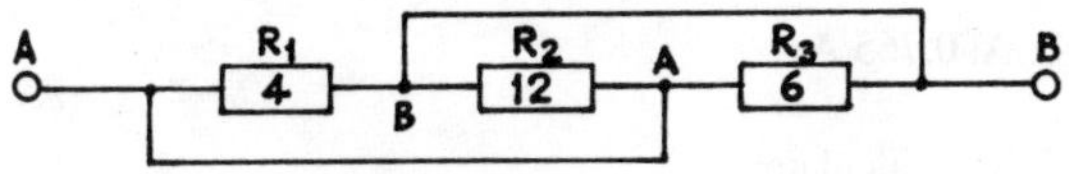

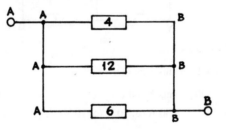

Fig. 69

R_1 has a terminal A and a terminal B

R_2 has a terminal A and a terminal B

R_3 has a terminal A and a terminal B

therefore the three resistors are re-drawn as shown in Fig. 70, showing three common terminals A and three common terminals B.

Fig. 70

$$R_{AB} = \frac{R_1 R_2 R_3}{R_1 R_2 + R_1 R_3 + R_2 R_3} = \frac{4 \times 12 \times 6}{4 \times 12 + 4 \times 6 + 12 \times 6} = 2\,\Omega$$

$$\text{or } \frac{1}{R_{AB}} = \frac{1}{4} + \frac{1}{12} + \frac{1}{6} = 0.25 + 0.0833 + 0.1666 = 0.5$$

$$\text{or } R_{AB} = \frac{1}{0.5} = 2\,\Omega.$$

SUMMARY 2

KIRCHHOFF'S CURRENT LAW (K.C.L.)

THE SUM OF THE CURRENTS FLOWING AT A JUNCTION OR NODE IS EQUAL TO ZERO

KIRCHHOFF'S VOLTAGE LAW (K.V.L.)

THE SUM OF THE EMF'S AROUND A CLOSED LOOP IS EQUAL TO ZERO.

EXERCISES 2

1. Using Kirchhoff's Laws calculate the current in each branch of the circuit of Fig. 71

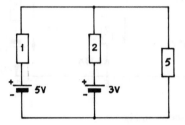

Fig. 71 Kirchhoff's Laws

(Ans.1.175 A, –0.41 A, 0.765 A)

2. Using Kirchhoff's Laws, calculate:

    (a) the three branch currents in the circuit of Fig. 72

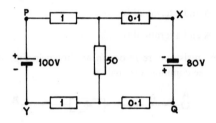

Fig. 72 Kirchhoff's Laws

(b) the p.d. between the points P and Q,

(c) the p.d. between the points X and Y.

(Ans. (a) 81.7 A, 1.27 A, 8.30 A (b) 10 V (c) 10 V)

3. Calculate the current in each circuit when the tapping is set such that AC = 25 Ω, in Fig. 73

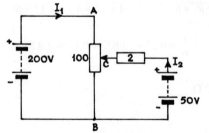

Fig. 73 Kirchhoff's Laws.

(Ans. 5.61 A, –4.82 A)

4. Apply Kirchhoff's current Law (K.C.L.) to the junction A and Kirchhoff's Voltage Law (K.V.L.) to the loops (1) and (2) of the circuit, of Fig. 74, and hence find I_1 and I_2.

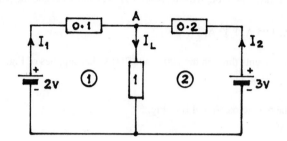

Fig. 74 Kirchhoff's Laws

(Ans. −1.87 A, 4.06 A)

5. Determine the currents I_1, I_2 and I_L in the circuit of Fig. 75.

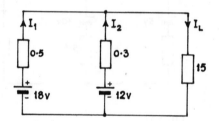

Fig. 75 Kirchhoff's Laws

(Ans. 7.86 A, $I_L = 0.95$ A)

6. Calculate the current in the network of Fig. 50.

(Ans. 5.83 A)

7. Referring to the circuit of example 10 Fig. 52, determine the currents I_1, I_2 and I_3 which are assumed to flow from the positive terminals or poles of the cells 2 V, 3 V and 5 V respectively. Redraw the circuit by inserting the actual flow of currents. State the discharging and charging currents.

(Ans. − 0.3125 A, 0.075 A, 0.2375 A)

8. Calculate the current through the load of 100 Ω in the circuit of Fig. 53.

(Ans. 4 A, 4.1 A)

9. Determine the magnitude and direction of the load current in the circuit of Fig. 54 and the p.d. across the load.

(Ans. 2.02 A, 1.99 A, 0.45 V)

10. Determine the actual currents through the 50 V, 20 V and 100 V d.c. supplies of Fig. 57 redraw the circuit indicating the actual currents flowing.

(Ans. −44 A, −14.8 A, 58.8 A)

11. Determine the currents through the resistors 1 Ω, 5 Ω and 15 Ω and the p.d. across A and B of Fig. 58.

   (Ans. 5.26 A, 3.95 A, 1.32 A, 19.73 V)

12. Calculate the currents through the 100 V and 200 V d.c. supplies of Fig. 59.

   (Ans. 48.5 A, −25.8 A)

13. Determine the p.d. across A and B of Fig. 59.

   (Ans. 22.14 V)

14. Determine the equivalent resistance between *A* and *G*, the diagonal of the cube or resistors of Fig. 58, employing Kirchhoff's Laws. (Ans. 5/6 Ω).

15. Two batteries of e.m.f s of 5 V and 10 V have internal resistances of 1 Ω and 5.5 Ω respectively are connected in parallel across a load of 4.5 Ω resistance. Determine the branch currents.

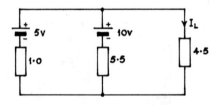

   (Ans. 0.144 A, 0.935 A, 1.079 A)

16. Two batteries of e.m.f s of 110 V and 120 V have internal resistances of 20 Ω and 40 Ω respectively are connected in parallel across a load of 100 Ω. Determine the p.d. across the load. (Ans. 100 V)

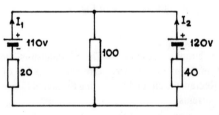

17. Two batteries of e.m.fs of 10 V and 5 V have internal resistances of 1 Ω and 1.5 Ω respectively are connected in parallel across a 10 Ω local. Determine the load current.

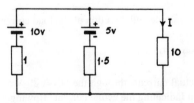

   (Ans. *I* = 1.75 A)

ELECTRIC FIELDS

3. Applies the fundamental laws and properties of electric fields to problems involving capacitors

   a. Introduces the concepts of electric field and electric flux to explain the forces of attraction and repulsion between charged bodies and defines electric field strength, potential and potential difference in terms of force and work done on a unit charge.

COULOMB'S LAW

The mechanical force, F, in newtons exerted between two charged bodies is given by the formula

$$F = \frac{Q_1 Q_2}{4\pi \varepsilon r^2}$$

where Q_1 and Q_2 are the charges of the bodies, r is the distance between the charges and ε is the permittivity of the medium.

If the charges are unlike, the force is *attractive* and if the charges are like, the force is *repulsive*. This formula describes Coulomb's law, the force is proportional to the product of the charges, and the force is inversely proportional to the square of their distance apart, r. If the medium is vacuum or free space then ε is ε_o and is the permittivity of free space ($\varepsilon_o = 8.85 \times 10^{-12}$ F/m). The force is reduced appreciably if the permittivity of the medium is high.

ELECTRIC FIELDS

A charged sphere has an electric field, that is, represented by a field of radial electric lines as shown in Fig. 76. Two parallel plates with positive charge on one plate has an electric field, represented by parallel lines as shown in Fig. 77.

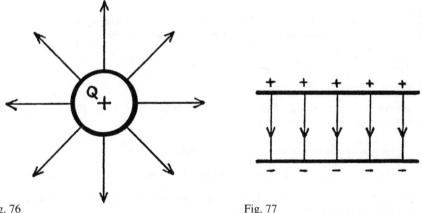

Fig. 76   Fig. 77

The region around the electric charge Q constitutes an electric field because any other charge placed in that region experiences a mechanical force given by Coulomb's law.

Therefore any region in which electric charges experience mechanical forces is called an electric field.

ELECTRIC FLUX

An electric field is represented by electric lines of force or electric flux. Electric flux emanates from a positive charged body as shown in Fig. 76.

Electric flux is equal to the electric charge Q coulombs.

$$Q = \text{electric flux (coulombs)}.$$

ELECTRIC FLUX DENSITY

The surface area of a sphere is given by $4\pi r^2$ where r is the radius of the sphere.

The electric flux, Q, that emanates in three dimensions from the sphere has an electric flux density

$$\boxed{D = \frac{Q}{A}} \qquad \text{... (2)}$$

D stands for displacement of electric flux density (coulombs per square metre) (C/m^2)
Q denotes electric flux in coulombs (C)
A denotes area in square metres (m^2).

ELECTRIC FIELD STRENGTH

The electric field strength is given in newtons per coulomb or volts per metre (V/m) and is denoted by the symbol E.

If the potential difference between the two parallel plates is V and the distance separated by the plates is d, then the electric field strength is given by

$$\boxed{E = \frac{V}{d}} \qquad \text{... (3)}$$

If a very small, positive point charge Q is placed at any point in an electric field and it experiences a force F, then the field strength E at that point is defined by the equation

$$\boxed{E = \frac{F}{Q}} \qquad \text{... (4)}$$

the electric field strength is the force in newtons per unit charge and its direction is that of the force F, thus the field strength E is a vector.

b. *Expresses field strength in terms of potential gradient.*

Potential gradient is the drop in potential per metre in the direction of the electric field.

The work done by a positive point charge of one coulomb in moving a short distance dx metres in the direction of the field is $E\,dx$ joules.

The potential difference in volts between two points is equal to the work done, in joules, in moving one coulomb of charge from one point to the other.

$$dv = E\,dx$$

the potential gradient is therefore given by E as

$$E = \frac{dv}{dx} \qquad \text{... (5)}$$

c. *Establishes the relationship between electric field strength and electric flux density and defines relative permittivity and permittivity of free space.*

The ratio of the electric flux density to the electric field strength gives the permittivity of the medium

$$E = \frac{D}{E} \qquad \text{... (6)}$$

D = electric flux density (coulombs per unit area)

E = electric field strength (volts per metre)

ε = absolute permittivity (farads per metre)

$\varepsilon = \varepsilon_o \varepsilon_r$ where ε_o = permittivity of free space

$$\varepsilon_o = 8.85 \times 10^{-12} \text{ F/M}$$

ε_r = relative permittivity (no dimensions)

$$\varepsilon_r = \frac{\varepsilon}{\varepsilon_o} \qquad \text{... (7)}$$

d. *Defines capacitance as the constant of proportionality between charge and potential difference and establishes the relationship between capacitance and the physical dimensions of a pair of parallel plates.*

THE CAPACITOR

Any two conductors between which an electric field can be maintained form a capacitor:

Consider two parallel plates as shown in Fig. 78 subject to a p.d. of V volts, one

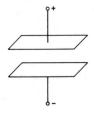

Fig. 78   Two parallel plate capacitor.

plate is positively charged and the opposite plate is negatively charged.

CAPACITANCE OF A PARALLEL PLATE CAPACITOR

The capacitance, $C = Q/V$ where Q is the charge in coulomb and V is the p.d. across the plates

$$D = \frac{Q}{A}$$

$$E = \frac{V}{d}$$

$$\frac{D}{E} = \varepsilon_o \varepsilon_r$$

$$C = \frac{Q}{V} = \frac{DA}{Ed} = \frac{\varepsilon_o \varepsilon_r A}{d}$$

$$\boxed{C = \varepsilon_o \varepsilon_r \frac{A}{d}} \qquad \dots (8)$$

For n plates of metal foil interleaved with a dielectric of mica or waxed paper, C is given by

$$\boxed{C = \frac{\varepsilon_o \varepsilon_r A (n-1)}{d}} \qquad \dots (9)$$

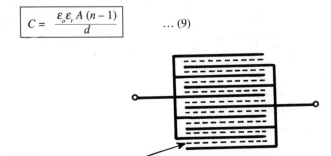

Fig. 79   Several parallel plate capacitor.

e.   Derives expressions for energy stored by a capacitor

A p.d. of V volts is applied across a capacitor C, the charge on the capacitor is given as

$$Q = CV$$

The charge Q is directly proportional to the p.d. applied, V

$$Q \propto V$$

The energy stored, W is equal to QV joules

$$W = QV.$$

Assuming that the capacitor is initially uncharged, the energy at that instant is zero. When V is applied, the energy stored is $W = QV$. The total energy stored in the capacitor is the average or mean value

$$W = \frac{O + QV}{2} = \frac{1}{2} QV$$

$$W = \frac{1}{2} QV$$

but $Q = CV$

$$W = \frac{1}{2} (CV) V = \frac{1}{2} CV^2$$

$$W = \frac{1}{2} CV^2 \qquad \dots (10)$$

the energy stored in a capacitor is thus proportional to the p.d. squared, therefore it is greatly dependent on the voltage.

f.   Solves problems relating to uniform fields in simple dielectrics involving the relationships established in a to e.

WORKED EXAMPLE 14

A 150 µF capacitor is connected across a 150 V d.c. supply. Determine (i) the charge (ii) the energy stored.

SOLUTION 14

(i) $Q = CV = 150 \times 10^{-6} \times 150 = 22.5$ mC

(ii) $W = \dfrac{1}{2} CV^2 = \dfrac{1}{2} \ 150 \times 10^{-6} \ 150^2 = 1.69$ J.

WORKED EXAMPLE 15

A p.d. of 100V d.c. is applied across the plates of a two-plate capacitor of capacitance 0.01 µF. The effective area of each plate is 100 mm^2 and the absolute permittivity of the dielectric is 150 pF/m

Determine: (i) the electric flux density

(ii) the electric field intensity

(iii) the relative permittivity

$\varepsilon_o = 8.85 \times 10^{-12}$ F/m.

SOLUTION 15

(i) $D = \dfrac{Q}{A} = \dfrac{CV}{A} = \dfrac{0.01 \times 10^6 \times 100}{100 \times 10^{-6}} = 0.01$ C/m^2

(ii) $E = \dfrac{D}{\varepsilon} = \dfrac{0.01}{150 \times 10^{-12}} = \dfrac{10^{10}}{150} = 6.67 \times 10^7$ V/m

(iii) $\varepsilon = \varepsilon_r \varepsilon_o$

$\varepsilon_r = \dfrac{\varepsilon}{\varepsilon_o} = \dfrac{150 \times 10^{-12}}{8.85 \times 10^{-12}} = 16.95.$

WORKED EXAMPLE 16

A two plate capacitor has an effective area of 75 cm^2 and the dielectric is air of 7.5 mm. A d.c. p.d. of 100 V is applied. Calculate:

(i)   the potential gradient in the dielectric

(ii) the capacitance

(iii) the electric flux

(iv) the electric flux density

$\varepsilon_o = 8.85 \times 10^{-12}$ F/m.

SOLUTION 16

(i) $E = \dfrac{V}{d} = \dfrac{100}{7.5 \times 10^{-3}} = 13.3$ KV/m

(ii) $C = \dfrac{\varepsilon_o \varepsilon_r A}{d} = \dfrac{8.85 \times 10^{-13} \times 1 \times 75 \times 10^{-4}}{7.5 \times 10^{-3}} = 8.85$ pF

(iii) $Q = CV = 8.85 \times 10^{-12} \times 100 = 8.85 \times 10^{-10}$ C

(iv) $D = \dfrac{Q}{A} = \dfrac{8.85 \times 10^{-10}}{75 \times 10^{-4}} = 0.118 \times 10^{-6}$ C/m².

WORKED EXAMPLE 17

A capacitor is formed by two parallel metal plates each of area 1,000 mm² and separated by a dielectric 0.5 mm thick. The capacitance of this arrangement is 500 pF and a d.c. p.d. of 20,000 V is applied to the terminals .

Calculate: (i) the charge on the plates

(ii) the relative permittivity of the dielectric

(iii) the electric flux density

(iv) the energy stored

$\varepsilon_o = 8.85 \times 10^{-12}$ F/m.

SOLUTION 17

(i) $Q = CV = 500 \times 10^{-12} \times 20,000 = 10$ μC

(ii) $\varepsilon = \dfrac{D}{E} = \dfrac{\dfrac{Q}{A}}{\dfrac{V}{d}} = \dfrac{\dfrac{10 \times 10^{-6}}{1000 \times 10^{-6}}}{\dfrac{20,000}{0.5 \times 10^{-3}}} = \dfrac{0.01}{4 \times 10^{7}}$

$\varepsilon = 0.25 \times 10^{-9}$ F/m

$\varepsilon_r = \dfrac{\varepsilon}{\varepsilon_o} = \dfrac{0.25 \times 10^{-9}}{8.85 \times 10^{-12}} = 28.2$

(iii) $D = \dfrac{Q}{A} = \dfrac{10 \times 10^{-6}}{1,000 \times 10^{-6}} = 0.01$ C/m²

(iv) $W = \dfrac{1}{2} CV^2 = \dfrac{1}{2} \times 500 \times 10^{-12} (20,000)^2 = 0.1$ J

g.   Deduces expressions for the equivalent capacitance of capacitors connected in series and parallel, solves simple problems and compares predicted and measured values for series and parallel capacitor combinations.

CAPACITORS IN SERIES

Consider two capacitors in series connected to a d.c. supply as shown in Fig. 80.

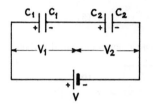

Fig. 80   Capacitors in series

Let V_1 and V_2 be the two p.d.s across C_1 and C_2 respectively.

The total charge of the circuit is also the charge on each capacitor since $Q = It$ and I is the same when the capacitors are charged

Using KVL $V = V_1 + V_2$ ... (1)

and since $Q = CV$ or $V = \dfrac{V}{C}$, $V_1 = \dfrac{Q}{C_1}$ and $V_2 = \dfrac{Q}{C_2}$

and substituting in (1), $\dfrac{Q}{C} = \dfrac{Q}{C_1} + \dfrac{Q}{C_2}$

dividing each term by Q, we have

$$\boxed{\frac{1}{C} = \frac{1}{C_1} + \frac{1}{C_2}}$$ ... (2)

If there are n capacitors in series, the formula is extended to

$$\boxed{\frac{1}{C} = \frac{1}{C_1} + \frac{1}{C_2} + \frac{1}{C_3} + ... + \frac{1}{C_n}}$$ ... (3)

CAPACITORS IN PARALLEL

The total charge of the system is shared between the two capacitors

$Q = Q_1 + Q_2$ ... (4)

Fig. 81 shows the circuit of two capacitors in parallel

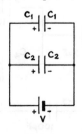

Fig. 81   Two capacitors in parallel

$$V = \frac{Q_1}{C_1} \ , \ V = \frac{Q_2}{C_2} \ , V = \frac{Q}{C}$$

Substituting these values in equation (4) we have

$$CV = C_1 V + C_2 V$$

dividing each term by V

$$\boxed{C = C_1 + C_2} \qquad \ldots (5)$$

The system can be extended to n capacitors in parallel, then the total capacitance

$$\boxed{C = C_1 + C_2 + C_3 + \ldots + C_n} \qquad \ldots (6)$$

WORKED EXAMPLE 18

Determine the equivalent capacitance of the capacitor networks of Fig. 82, 83, 84, 85:

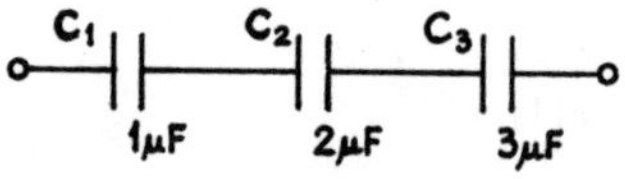

Fig. 82   Capacitors in series

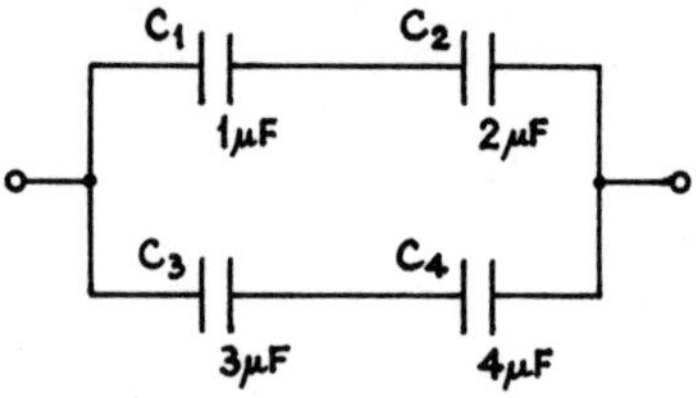

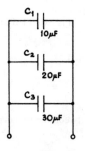

Fig. 83   Capacitors in series and parallel

Fig. 84   Capacitors in parallel

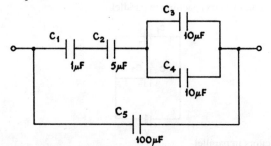

Fig. 85   Capacitors in series and parallel

SOLUTION 18

(i) Referring to Fig. 82

$$\frac{1}{C} = \frac{1}{C_1} + \frac{1}{C_2} + \frac{1}{C_3} = \left(\frac{1}{1} + \frac{1}{2} + \frac{1}{3}\right) 10^6 = \frac{6+3+2}{6} = \frac{11}{6} \times 10^6$$

therefore $C = \frac{6}{11}$ μF

or $\quad C = \dfrac{C_1 C_2 C_3}{C_1 C_2 + C_1 C_3 + C_2 C_3} = \dfrac{1 \times 2 \times 3}{1 \times 2 + 1 \times 3 + 2 \times 3} \times 10^{-6} = \dfrac{6}{11}$ μF

(ii) $\quad C = C_{T_1} + C_{T_2} = \dfrac{C_1 C_2}{C_1 + C_2} + \dfrac{C_3 C_4}{C_3 + C_4} = \left(\dfrac{1 \times 2}{1+2} + \dfrac{3 \times 4}{3+4}\right)$ μF $= \left(\dfrac{2}{3} + \dfrac{12}{7}\right)$ μF $= \dfrac{50}{21}$ μF

(iii) Referring to Fig. 83

$$C = C_1 + C_2 + C_3 = (10 + 20 + 30)\ \mu F = 60\ \mu F$$

(iv) $\quad C_{T_1} = \dfrac{C_1 C_2}{C_1 + C_2} = \dfrac{1 \times 5}{1+5}$ μF $= \dfrac{5}{6}$ μF

$$C_{T_2} = C_3 + C_4 = (10 + 10)\ \mu F = 20\ \mu F$$

$$C = \dfrac{C_{T_1} C_{T_2}}{C_{T_1} + C_{T_2}} = \dfrac{(5/6)\,20}{5/6 + 20} = 0.8\ \mu F$$

$$C_T + (0.8 + 100)\ \mu F = 100.8\ \mu F$$

WORKED EXAMPLE 19

For the circuit shown in Fig. 86 calculate:

(i)   the equivalent capacitance

(ii) the p.d. across the 30 μF

(iii) the total energy stored.

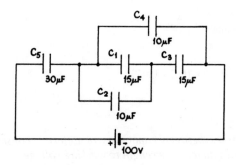

Fig. 86   Capacitors in series and parallel

SOLUTION 19

(i)   The equivalent capacitance can be found by first taking C_1 and C_2 in parallel $C_{T_1} = C_1 + C_2$,

then taking the two capacitors C_{T_1} and C_3 in series, $\dfrac{C_{T_1}C_3}{C_T + C_3} = C_T$, then taking C_T and C_4

in parallel, $C_T + C_4 = C_{T_2}$ and finally $C = \dfrac{C_5 C_{T_2}}{C_5 + C_{T_2}}$, the equivalent capacitance

$$C_{T_1} = C_1 + C_2 = (15 + 10)\ \mu F = 25\ \mu F$$

$$C_T = \frac{C_{T_1}C_3}{C_{T_1} + C_3} = \frac{25 \times 15}{25 + 15}\ \mu F = 9.375\ \mu F$$

$$C_{T_2} = C_T + C_4 = (9.375 + 10)\ \mu F = 19.375\ \mu F$$

$$C = \frac{C_5 C_{T_2}}{C_5 + C_T} = \frac{30 \times 19.375}{30 + 19.375} = 11.77\ \mu F.$$

(ii) In order to find the p.d. across 30 μF capacitor, we have to consider Fig. 87 where C_5 and C_{T_2} are in series across the 100 V d.c. supply

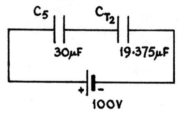

Fig. 87   Equivalent circuit of capacitors

The total charge of the system is found from the formula $Q = CV = 11.77 \times 100 \times 10^{-6}$

$$Q = 1177\ \mu C\ \text{or}\ 1.18\ \mu C$$

The p.d. across 30 μF is found from $V = A/C_5$

$$V = \frac{1.177 \times 10^{-3}}{30 \times 10^{-6}} = 39.2\ \text{volts}$$

(iii) $W = \dfrac{1}{2}\ CV^2 = \dfrac{1}{2} \times 11.77 \times 10^{-6} \times 100^2 = 0.059\ J$

$W = 59\ \text{mJ}$

h.   Identifies and distinguishes between capacitors of differing construction and characteristics and relates dielectric strength to working voltage.

Most practical capacitors have their plates separated by an insulating material which is called a dielectric. It was shown earlier that the effect of introducing a dielectric was to increase the capacitance of the capacitor, according to the formula

$$C = \frac{\varepsilon_o \varepsilon_r A}{d}$$

where ε_r is the relative permittivity of the dielectric constant.

It is important that the dielectric shall be able to withstand the potential gradient applied on it when the capacitor is charged.

The resistance of the dielectric layer is Known as *the insulation resistance* which must be extremely high for a good component, in the order G Ω. The insulation resistance depends on the area and thickness of the dielectric and falls with increase in temperature.

For a large capacitance, the thickness of a dielectric is thin and the insulation resistance therefore will be low.

If the potential gradient applied to a capacitor is exceeded from the quoted by the manufacturer, the insulation is likely to break down.

A capacitor is rated to withstand a maximum *working voltage*.

Different dielectric materials differ in the potential gradient they can withstand and to test a particular material, a dielectric of 1mm thick is placed between the two electrodes, and the p.d. is then increased gradually across the plates until the insulation resistance breaks down. The potential gradient sufficient to cause breakdown of the insulation is called the *dielectric strength* of the material and is expressed in KV per mm. The dielectric strength in KV/mm for some materials is given.

Material	Dielectric Strength KV/mm
Glass	20
Mica	40 to 100
Polythene	40
Barium Titanate	4
Air	1
Oil (Insulating)	40

SUMMARY 3

$$F = \frac{Q_1 Q_2}{4\pi\varepsilon r^2} \quad (N)$$

$$\varepsilon = \varepsilon_r \varepsilon_o \quad (F/m)$$

$$D = \frac{Q}{A} \quad (C/m^2)$$

$$C = \frac{\varepsilon_o \varepsilon_r A}{d} \quad (F)$$

$$E = \frac{V}{d} \quad (V/m)$$

$$E = \frac{F}{Q} \quad (N/C)$$

$$C = \frac{\varepsilon_o \varepsilon_r A (n-1)}{d} \quad (F)$$

$$E = \frac{dV}{dx} \text{ (V/m)} \qquad\qquad\qquad Q = CV$$

$$\varepsilon = \frac{D}{E} \text{ (F/m)} \qquad\qquad\qquad W = \frac{1}{2}CV^2 \text{ (J)}$$

Capacitors in series $\quad \dfrac{1}{C} = \dfrac{1}{C_1} + \dfrac{1}{C_2} + \dots + \dfrac{1}{C_n}$

Capacitors in parallel $C = C_1 + C_2 + \dots + C_n$.

EXERCISES 3

1. A parallel plate capacitor has a capacitance of 100 pF. What will be the new value of capacitance if the effective plate area is reduced by 50% and the dielectric thickness is doubled?

    (Ans. 25 pF).

2. (a) What factors determine the capacitance existing between a parallel metal plates?

    (b) Write down the formula for the capacitance connecting the above factors and state the unit of each physical quantity.

    (c) For a given capacitor write down the expressions stating the units employed relating:

    (i) the potential difference and charge

    (ii) the potential difference and stored energy.

3. Two air-spaced metal plates insulated from each other are mounted so as to form a parallel plate capacitor. When the distance between the plates is 2.5 mm the capacitance is 5 pF. Determine the capacitance when the spacing is increased to 3.5 mm.

    How will the value of the original capacitor be affected if the whole is immersed in oil having a relative permittivity of 5?

    (Ans. 3.57 pF, 25 pF)

4. A p.d. of 15 KV is applied to the terminals of a capacitor consisting of two circular plates, each having an area of 100 cm^2, separated by a dielectric 2.5 mm thick. If the capacitance is 100 pF, calculate:

    (i) the electric flux

    (ii) the electric flux density

    (iii) the relative permittivity of the dielectric ($\varepsilon_o = 8.85 \times 10^{-12}$ F/m).

    (Ans. (i) 1.5 μC (ii) 1.5×10^{-4} C/m^2 (iii) 2.82)

5. A capacitor is charged with 100 μC, if the energy stored is 1 μJ find (i) the p.d. (ii) the value of the capacitance.

    (Ans. (i) 0.02 V (ii) 5,000 μF)

6. Two capacitors A and B have capacitances of 10 μF and 20 μF respectively.

Calculate the voltage across A and the energy stored in A when A and B are connected (i) in series and (ii) in parallel with a d.c. supply of 20 V in both cases.

(Ans. (i) 13.3 V, 888 μJ (ii) 20 V, 2 mJ)

7. Two capacitors 10 μF and 30 μF are connected in series. Across both of these is connected a capacitor of 20 μF. The whole combination is connected across a 100 V d.c. supply.

Calculate: (i) the equivalent capacitance

(ii) the p.d. across the 30 μF capacitor

(iii) the charge on 10 μF capacitor

(iv) the energy stored by the 20 μF capacitor.

(Ans. (i) 27.5 mF (ii) 750 mC (iii) 750 mC (iv) 0.1 J)

8. A 1 μF capacitor charged to a p.d. of 100 V is connected across a second uncharged capacitor of 2 μF, the source of voltage being removed before the second capacitor is added to the first.

Find (i) the resulting p.d. across the combined capacitors

(ii) the energy stored in each capacitor before and after paralleling. Comment on the results obtained.

(iii) if the two capacitors were discharged and then connected in series across the 100V supply what would be the voltage across each capacitor?

(Ans. (i) 33.3 V (ii) 0.005 J, 0 J; 554 μJ, 1,109 μJ, 5 mJ, 1.66 mJ, Loss 3.34 mJ (iii) 66.7 V, 33.3 V)

9. Calculate: (i) the charge on each capacitor before C is connected.

(ii) the total energy stored

(iii) the value of C which when connected in parallel with the 20 μF capacitor would change V_2, to 60 V with the same supply.

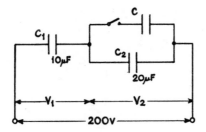

Fig. 88   Series-parallel capacitors sharing charge.

(Ans. (i) 1.33 mC (ii) 0.132 J, 140 V, 60 V (iii) 3.3 μF)

10. Determine the total value of capacitance for each capacitor system in Fig. 89 and Fig. 90

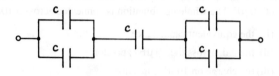

Fig. 89   Series-parallel capacitors

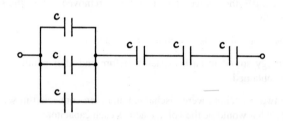

Fig. 90   Series-parallel capacitors

When C is (a) 1 μF and C is (b) 10 μF.

(Ans. (a) 0.5 μF, 5 μF (b) $\frac{3}{10}$ μF, 3 μF)

MAGNETIC FIELDS

4. *Applies the fundamental laws governing magnetic fields to the solution of problems relating to magnetic circuits and materials.*
   a. *Introduces the concept of the magnetic field and magnetic flux to explain the forces of attraction and repulsion between magnetised bodies and defines magnetic field strength.*

MAGNETIC FIELD

A number of magnetic lines make up a magnetic field. A bar magnet or a permanent magnet possesses a magnetic field which can be detected by a small compass needle or by sprinkling iron filings in the field. The field due to a bar magnet is illustrated in Fig. 91, while the magnetic field due to a solenoid is shown in Fig. 92.

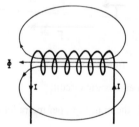

Fig. 91   The magnetic field round a bar magnet

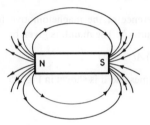

Fig. 92   The magnetic field due to a solenoid

MAGNETIC FLUX

The magnetic lines shown are flowing from the north pole, N and into the south pole. The magnetic lines *do not* intersect.

The magnetic lines of force or the magnetic flux is denoted by the Greek letter φ and it is expressed in webers (Wb). Faraday introduced the lines of flux as a pictorial method of representing the distribution of a magnetic field and they are used to define field quantitatively.

The magnetic fields of two like poles and two unlike poles are shown in Fig. 93 and Fig. 94.

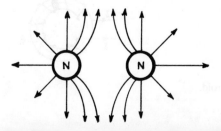

Fig. 93   Like magnetic poles

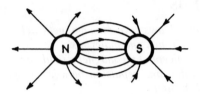

Fig. 94   Unlike magnetic poles

In Fig. 93 the magnetic field between the like poles is weak, and in Fig. 94 the magnetic field between the unlike poles is strong. The former field exhibits a force of repulsion and the latter field exhibits a force of attraction.

MAGNETIC FIELD STRENGTH OR MAGNETISING FORCE

The magnetic potential difference or the magnetomotive force (m.m.f.) is produced by the current flow in a number of turns N. The m.m.f. is given by

$F = IN$ (ampere-turns) At

The same current is used N times and it is shown in Fig. 95

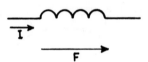

Fig. 95   The magnetic potential difference (m.m.f.)

This is analogous to the e.m.f. in an electric circuit.

The magnetic field strength, H, is the m.m.f., F, per metre length, l, of the flux path

$$H = \frac{F}{l} = \frac{IN}{l} \ (At/m)$$

where l is the length of the flux path or the magnetic length.

WORKED EXAMPLE 20

A toroid shown in Fig. 96 is wound uniformly with 1000 turns and carries a current $I = 2$ A, the mean diameter of the toroid is 30 cm. Determine the magnetic field strength.

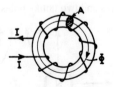

Fig. 96   The toroid.

SOLUTION 20

The magnetic flux, ϕ, flows in the direction shown according to the corkscrew rule and the magnetic length is πD, the flux path

$$H = \frac{IN}{l} = \frac{2 \times 1000}{\pi \, 0.3} = 2{,}122 \text{ At/m}$$

b. *Establishes the relationship between magnetic field strength and magnetic flux density and defines relative permeability and the permeability of free space.*

MAGNETIC FLUX DENSITY

The magnetic line of flux ϕ webers which cross an area A m^2 is termed as magnetic flux density, B

$$B = \frac{\phi}{A} \quad \text{(tesla or Wb/m}^2\text{)}$$

In Fig. 97, the magnetic flux, ϕ, crosses the cross sectional area, A, to give the flux density.

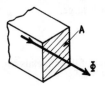

Fig. 97 Flux density

PERMEABILITY

The relationship between magnetic field strength and magnetic flux density is given by the equation

$$\frac{B}{H} = \mu$$

and $\mu = \mu_o \mu_r$.

The ratio of B to H is termed as absolute permeability μ where μ_o is the permeability of free space which is a constant given as $4\pi \, 10^{-7}$ H/M (henrys per metre) and μ_r is the relative permeability which is a number and has no dimensions ($\mu_r = 1$ for a non-magnetic material).

The permeability is a measure of the magnetic performance of a material.

c. *Investigates the effect of the core material on the performance of an electromagnet, compares magnetisation characteristics of typical ferromagnetic materials obtained by measurement and deduces the range of values of relative permeability.*

AN ELECTROMAGNET,

A number of turns are wound on a ferromagnetic core as shown in Fig. 98.

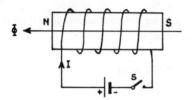

Fig. 98   The electromagnet.

When the switch, S, is closed, current I flows in the windings thus setting up a magnetic as shown of flux density ϕ. This is the principle of an electromagnet. Its advantages over the permanent bar magnet is that it may be switched on and off as required. It is a temporary magnet and must be made of soft iron.

Referring to the previous formular.

$$B = \frac{\phi}{A}, \qquad H = \frac{IN}{l} \qquad \text{and} \qquad \frac{B}{H} = \mu = \mu_o \, \mu_r.$$

If the core is non-magnetic then $\mu_r = 1$ and the ratio B/H is minimum. If the core is ferromagnetic then μ_r is large (e.g. 7,000) and B/H is large.

MAGNETIC MATERIALS

Iron, nickel, cobalt and certain alloys of these metals can be made into strong magnets, and are known as *ferromagnetic* materials.

Steel is an alloy of iron, made by adding small percentage of carbon to pure iron and it is a much harder metal than pure iron.

A permanent magnet can be made by placing a steel bar or rod inside a solenoid and the current switched on. The steel becomes a magnet and can pick up pieces of certain metals.

Alloys of nickel and cobalt are used for making powerful permanent magnets and are called ferrites. These magnetize strongly and are not easily demagnetized. Ferrites have a high electrical resistance, and hence the current and power loss are extremely small.

The graph of B against H for magnetic and non-magnetic material is shown in Fig. 99.

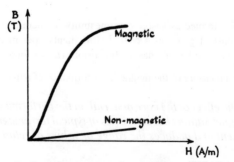

Fig. 99   B/H graphs for magnetic and non-magnetic materials.

The magnetization curve for a magnetic material is non-linear and for a non-magnetic material is linear. The former has a variable permeability and the latter a constant permeability, this is shown in Fig. 100 where μ_r is plotted against H.

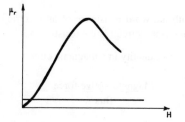

Fig. 100 μ_r/H graphs for magnetic and non-magnetic.

There are 'soft' magnetic materials and 'hard' magnetic materials. The former have high saturation B and low loss (iron, steel, silicon steel, nickel-iron alloys $(30 - 70\% N_i)$, cobalt - iron alloys, manganese and aluminium oxides (ferrites) and the latter have high residual magnetic, effect. The BH curves for the soft and hard magnetic materials are shown in Fig. 101

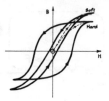

Fig. 101 BH curves for soft and hard materials.

Magnetisation curves for silicon iron, mild steel, and cast iron are shown in Fig. 102. The relative permeability plotted against the flux density for these materials are shown in Fig. 103.

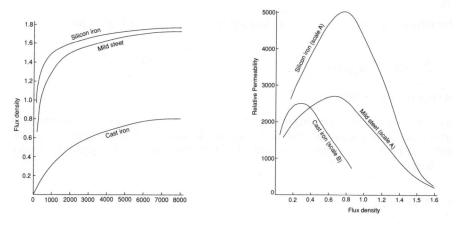

Fig. 102 BH curves for various materials.

Fig. 103 Curves of relative permeability against flux density for various materials.

It can be deduced from the graphs of μ_r/B that values of $\mu_r = 5000$ for silicon iron at $B = 0.75$ T, $\mu_r = 2,500$ for cast iron at $B = 0.225$ T and $\mu_r = 2600$ for mild steel at $B = 0.625$ T may be obtained.

d. *Relates magnetomotive force, reluctance and magnetic field strength and solves problems involving magnetic circuits having not more than a single change of dimension, material or air-gap using data from magnetisation curves.*

RELUCTANCE

The student is familiar with the word resistance of an electric circuit, that is the controlling component of current in an electric circuit is the resistor.

Reluctance, S, is the analogous quantity in a magnetic circuit

$$\text{Reluctance} = S = \frac{F}{\phi} = \frac{\text{Magnetomotive force}}{\text{flux}} \qquad ...(1)$$

and it is expressed in A/Wb (amperes/weber). Reluctance is analogous to resistance in an electric circuit. The formula of equation (1) is analogous to the formula of resistance, $R = V/I$ which is ohm's law.

Fig. 104 which indicates the property of the reluctance which is the magnetic resistance of the magnetic circuit.

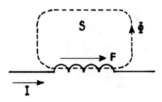

Fig. 104   Reluctance

$$S = \frac{F}{\phi} = \frac{F}{BA} = \frac{F}{\mu HA} . \qquad ...(2)$$

The relationship between m.m.f., reluctance and magnetic field strength, H is given

$$S = \frac{F}{\mu HA} = \frac{F}{\mu_o \mu_r HA} . \qquad ...(3)$$

WORKED EXAMPLE 21

A coil of 500 turns, wound on ring of magnetic material, carries a current of 1A. The length and cross-sectional area of the flux path in the ring are 500 mm and 1000 mm² respectively. The realtive permeability, $\mu_r = 3500$ and $\mu_o = 4\pi \times 10^{-7}$ H/m. Determine (i) the magnetic flux density (ii) the reluctance.

SOLUTION 21

$$S = \frac{E}{\phi} = \frac{IN}{BA} \qquad\qquad \text{and} \quad \frac{B}{H} = \mu_o \mu_r = \frac{B}{I\,N/l} = \frac{Bl}{IN}$$

(i) $B = \mu_o \mu_r \dfrac{IN}{l} = \dfrac{4\pi \times 10^{-7} \times 3,500 \times 1 \times 500}{0.5} = 4.4\ \text{T}$

(ii) $S = \dfrac{IN}{BA} = \dfrac{1 \times 500}{4.4 \times 10^{-3}} = 113,636\ \text{A/Wb}$

WORKED EXAMPLE 22

A piece of iron has a magnetic length of 150 mm a cross-sectional area of 500 mm², and a relative permeability of 550. Find the reluctance of the iron. ($\mu_o = 4\pi \times 10^{-7}$ H/m)

SOLUTION 22

$$S = \frac{F}{\phi} = \frac{IN}{BA} = \frac{IN}{H\mu_o \mu_r A} = \frac{IN \, l}{IN \, \mu_o \mu_r A} = \frac{l}{\mu_o \mu_r A}$$

$$S = \frac{l}{\mu_o \mu_r A} = \frac{0.15}{4\pi \times 10^{-7} \times 550 \times 500 \times 10^{-6}} = 434,059 \text{ A/Wb.}$$

AIR GAP

A toroid of a magnetic material has a radial cut as shown in Fig. 105.

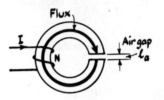

Fig. 105   A radial cut in a toroid.

The magnetic flux set up in the core is shown which also passes through the air gap.

The magnetic circuit is now made up of two reluctances in series since the magnetic flux is the same in the core and in the air gap, assuming no magnetic leakage occurs.

The magnetic lengths are l_i that of the iron, and l_a, that of the air gap.

The total reluctance of the magnetic circuit is given by

$$S = S_i + S_a = \frac{l_i}{\mu_o \mu_r A} + \frac{l_a}{\mu_o A}$$

since $\mu_r = 1$ for the air gap.

It is obvious that the effect of the air gap is to increase the reluctance of the magnetic circuit. This argument is analogous to that of one electric circuit, the total resistance is equal to the sum of the resistances since the same current flows through each resistor.

COMPARISON BETWEEN ELECTRIC AND MAGNETIC CIRCUIT

ELECTRIC CIRCUITS			MAGNETIC CIRCUITS		
TERM	SYMBOL	UNIT	TERM	SYMBOL	UNIT
Current	I	A	Flux	ϕ	Wb
Electromotive force	E	V	Magnetomotive force	F	A
Current Density	J	A/m²	Flux density	B	Wb/m² (T)
Resistance	R	Ω	Reluctance	S	A/Wb
Conductivity	σ	S/m	Permeability	μ	H/m
Resistivity	ρ	Ωm		$1/\mu$	m/H

PERMEABILITY IS ANALOGOUS TO CONDUCTIVITY

$$R = \rho \, \frac{l}{A} \quad \text{or} \quad \rho = \frac{RA}{l} \quad \text{or} \quad \sigma = \frac{1}{\rho} = \frac{l}{RA} = \text{conductivity}$$

$$\mu = \frac{B}{H} = \frac{\phi/A}{F/l} = \frac{\phi l}{FA} = \frac{l}{SA} \quad \text{permeability}$$

$$S = \frac{l}{\mu A} = \left(\frac{1}{\mu}\right) \frac{l}{A} \quad \text{analogous to } R = (\rho) \, \frac{l}{A}$$

therefore $\dfrac{1}{\mu}$ is analogous to ρ or μ is anologous to σ.

WORKED EXAMPLE 23

A 1500 turn coil is wound on an iron ring having a c.s.a. of 10 cm², and a mean diameter of 0.15 m. There is a radial saw cut 1 mm wide in the ring.

Calculate the current in the coil for a flux of 1 mWb in the ring. The B/H curve for the iron is as follows:

B (T)	0.62	0.88	1.03	1.13	1.18
H (A/m)	150	250	350	450	550

$\mu_o = 4\pi \times 10^{-7}$ H/m

SOLUTION 23

The magnetisation curve B/H is plotted on Fig. 106.

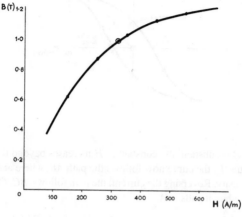

Fig. 106   *B/H* Curve

$$B = \phi/A = \frac{1 \times 10^{-3}}{10 \times 10^{-4}} = 1 \text{ T}$$

From the graph of Fig. 106, $H_i = 320$ A/m corresponding to $B = 1$ T. If the air gap is not taken into account then $H_i = 320 = I_i N/0.471$

$$H_i \quad \frac{I_i\,1500}{0.471} \quad \text{or } 1500\,I_i = 0.471 \times 320$$

current due to the iron, $I_i = \dfrac{320 \times 0.471}{1500} = 0.1\text{A}$

$$H_a = \frac{S_a\,\phi}{l_a} = \frac{l_a}{\mu A}\,\frac{\phi}{l_a} = \frac{\phi}{\mu_o A} = \frac{1 \times 10^{-3}}{4\pi \times 10^{-7} \times 10 \times 10^{-4}} = 795,775 \text{ A/m.}$$

The total magnetic field intensity

$$H = 320 + 795,775 = 796,095 \text{ A/m}$$

Current due to the air gap

$$I_a = \frac{H_a l_a}{N} = \frac{795775 \times 0.001}{1500} = 0.531 \text{ A}$$

The total current = 0.1 + 0.531 = 0.631 A.

e.   *Displays hysteresis loops on a C.R.O. and observes the effects of variation of magnetic material and magnetic field strength.*

HYSTERESIS

Hysteresis is a Greek word meaning 'lagging' and in this context, means that the flux density, B, is lagging behind the magnetic field intensity or magnetising force, H, when a specimen of ferromagnetic material is taken through a cycle of magnetisation.

The specimen is initially in a demagnetised state. As H increases, the magnetisation curve is non-linear as we saw previously and is shown as OA in Fig. 107.

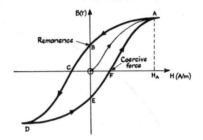

Fig. 107   Hysteresis loop

At A it saturates, that is, B is substantially constant as H increases beyond the point A. If H or I is decreased from its value H_A, the curve now follows the path AB, where at B, $OB = B_r$ when H = OB_r = remanent flux density. Reversing the current, the path follows, BC, CD, DE, EF and one complete cycle is traced when it ends at A.

$OF = H_c$ = coersive force, when B = zero tesla. The closed loop $ABCDEFA$, is called a hysteresis loop when H has been reduced from H_A to zero the iron is still magnetised in the positive direction and the magnetic flux density at that point is called the *remanent flux density*, since there is a residual magnetism in the iron. The amount of negative magnetising force which must be applied in order to annul the remanent flux density is called *the coersive force*.

The remanent flux density corresponding to saturation of the material is called *remanence* and the corresponding coercive force is called coercivity.

f.   States that hysteresis loop is proportional to the area of the loop and explains the importance of hysteresis loss for a.c. excited devices.

The energy stored in the magnetic material is proportional to the area of the loop. If can be shown that the energy expended per cycle per cubic metre is equal to the area of the hysteresis loop in joules.

Steinmitz devised an empirical formula, namely

Energy dissipated $= \eta B^{1.6}$ joules per m^3 per cycle

where η is the hysteresis coefficient and has numerical value between 250 and 300 for 4% silicon steel, 500 for sheet steel, 750 – 3000 for cost iron steel and 3,000 – 4000 for cost iron.

Energy dissipated per second $= \eta f B^{1.6}$ in joules per m^3 where f is the frequency. Therefore the power loss $= \eta f B_m^{1.6}$ W/m^3.

SUMMARY 4

$F = In$

$H = IN/l$

$B = \phi/A$

$B/H = \mu$

$\mu = \mu_o \mu_r$

$S = F/\phi$

$S = l/\mu_o \mu_r A$

$S_a = S_i + S_a = \dfrac{l_i}{\mu_o \mu_r A} + \dfrac{l_a}{\mu_o A}$

for series magnetic circuits

EXERCISE 4

1. A ring specimen of cost steel has a mean length of 50 cm and a cross sectional area of 10 cm^2.

    If 300 turns of wire are wound uniformly round the ring, calculate the current required in these turns to set up a magnetic flux of 750 μ Wb in ring and gap. (μ_r = 1000 for cost steel and μ_o = 4π × 10^{-7} H/m (Ans. 1 A)

2. A perspex toroid has a mean circumference of 75 cm and a c.s.a of 10 cm^2. The oil consists of 500 turns and carries a current of 5 A.

    Calculate : (i) the magnetomotive force

    (ii) the magnetic field intensity

    (iii) the flux density

    (iv) the total magnetic flux

    (Ans. (i) 2,500 A (ii) 3,333 A/m (iii) 4.19 m T (iv) 4.19 μ Wb)

3. A coil having 1000 turns is wound on an iron core having a mean length of 100 cm and c.s.a of 7.5 cm^2. Calculate the current required to produce a flux-density if 1.5 T in the core.

    What would be the value of the flux-density of there was an equavalent of a 1 mm air gap in the ring and current remained the same.

    μ_r = 1000, and μ_o = 4π × 10^{-7} H/m.

    (Ans. 1.19 A 1,591 A/Wb, 997 T)

4. (a) A ring speciment made of steel-alloy has a mean diameter of 20 cm and a cross-sectional area of 6 cm^2. In a certain magnetic test the reluctance of the ring is found to be 400,000 A/Wb. calculate the relative permeability of the steel-alloy in this case.

    (b) A radial airgap is now made in the ring and it is uniformly wound with a coil of 250 turns. When a current of 1 A flows in the coil, it is found that the reluctance of the air gap is the same as that for the ring, which is the same as that in (a). Calculate the magnetic flux set up in the ring.

    (Ans. (a) 1.25, (b) 313 μ Wb)

5. (a) Make a neat sketch of a hysteresis loop and with its aid explain the terms:

    (b) (i) remanent flux density, B_r,

    (ii) Coersive force, H_c,

    (iii) initial magnetisation,

    (iv) magnetic saturation.

    (b) A ring specimen of cost steel has radial air gap 0.5 mm long cut through the ring and the air gap has a reluctance of 5 × 10^5 A/Wb. The ring has a mean length of 50 cm and relative permeability of 1000. Calculate the number of turns wound on the ring if 1.5 A is required in these turns to set up a flux of 500 mWb

    (Ans. (a) (b) 398 A/Wb, 500,398 A/Wb, N = 167)

6. A ferromagnetic toroid has a mean circumference of 25 cm, a cross sectional area of 2 cm² and is wound with a coil of 1000 turns.

   A 1 mm radial air gap is cut in the toroid. Calculate the coil current required to establish a flux of 100 μ Wb in the toroid.

   A magnetisation curve for the specimen is given: B (T)   0.47   0.60   0.66

                                                 H (A/m)   2,000   3,000   4,000

   (Ans. 192 μA)

7. (i) Write down an equation for the resistance of a copper wire in terms of its length and cross sectional area, giving the units for each quantity. Give the analogous expression in a magnetic circuit and explain the term of each quantity and the respective units.

   (ii) Write down an equation for the magnetomotive force in terms of magnetic flux, magnetic length, permeability and cross sectional area. Give the analogous expression in an electric circuit.

8. A ferro magnetic toroid ring with a mean total length round the iron of 400 mm, the cross-section being 400 mm² the number of turns wrapped round uniformly is 100 and the direct current is 5 A. Calculate:

   (i) the reluctance of the ring

   (ii) the flux in the iron

   $\mu_o \mu_r = 4\pi \times 10^{-4}$ H/m

   (Ans. (i) 795,775 A/Wb, (ii) 628 mWb)

9. In question 8 an air gap of 4 mm is cut calculate the reluctance in the air gap what is the total reluctance of the circuit? What is the effect of cutting an air gap?

   (Ans. 8,761, 560 A/Wb)

ELECTROMAGNETISM

5. *Applies the fundamental principles and laws governing electromagnetic induction.*
   a. *Explains the motor principles in terms of the interaction between a magnetic field and a current - carrying conductor and applies the relationship F = Bli to simple situations.*

THE MAGNETIC EFFECT OF CURRENT

Consider a current-carrying conductor as shown in Fig. 108. A magnetic field is set up around the conductor according to the *corkscrew* rule as shown in the diagram or the right hand gripping rule.

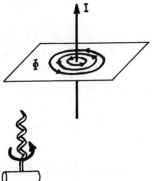

Fig. 108   The magnetic field around a current carrying conductor.

Fig. 109 shows the current flowing out of the paper and Fig. 110 shows the current flowing into the paper.

Fig. 109   The current flowing out of the paper.

Fig. 110   The current flowing into the paper.

The force on the conductor, F, in newtons depends directly on the flux density, B, the length of the conductor, l and the current flowing through the conductor, I,

$$F = BlI.$$

If the current $I = 1$ A, $l = 1$ m, $B = 1$ T then the force on the conductor $F = 1$ newton. If the conductor makes an angle θ with the magnetic field then F is given

$$F = BlI \sin \theta.$$

MOTOR PRINCIPLE

The left hand rule illustrates clearly the *motor principle* as shown in Fig. 111.

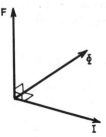

Fig. 111   The Left Hand Rule.

The thumb, the first finger and the middle finger of the left hand when they are stretched out at right angles mutually, we have the force along the thumb, the magnetic field *B* along the first finger and the current along the middle finger.

If a current – carrying conductor is placed at right angles to a magnetic field as shown in the Fig. 112, the conductor

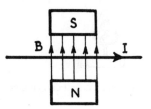

Fig. 112   Magnetic field being by a current-carrying conductor.

is moved upwards out of the paper with a force $F = BlI$, where B is the flux density set up between the north and south, l is the length of the conductor in the magnetic field, and I is the current flowing through the conductor. The current I generates a force, F, out of the paper. This is the *motor principle*. If the current is reversed then the force acts into paper. Fig. 113 and Fig. 114 show the forces into and out of the paper respectively.

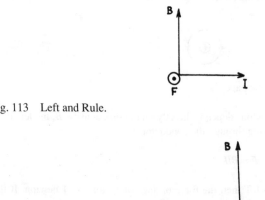

Fig. 113   Left and Rule.

Fig. 114   Right Hand Rule.

WORKED EXAMPLE 24

A conductor 45 cm in length and carrying a current of 0.3 A is placed at right angles to a magnetic field having a flux density of 1.0 T. Calculate the force on the conductor.

SOLUTION 24

$F = BIl$

where B = flux density in teslas (T)

l = length of conductor (m)

I = current in amperes (A)

F = force in newtons (N)

$F = 1 \times 0.3 \times 0.45 = 0.135$ N.

b. *Observes the production of induced voltage and describes it as due to change in flux linkage.*

Consider the simple situation shown in Fig. 115

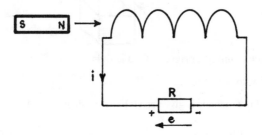

Fig. 115   Induced e.m.f. Lenz's Law.

A stationary coil is connected to a resistor for continuity. A permanent magnet is moved along the arrow shown towards the coil. The flux linkage, $N \phi$ is changing whilst the magnet is moving. Due to this change in flux linkage, an induced e.m.f. e, is established across R as shown. If the magnet moves now in the opposite direction as hown in Fig. 116, the induced

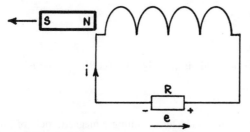

Fig. 116   Induced e.m.f. Lenz's Law.

e.m.f. is reversed. The induced e.m.f. is directly proportional to the rate of change of flux linkage.

$$e = \frac{d}{dt} (N\phi)$$

Since N is the number of turns of the coil which is constant

$$e = N \frac{d\phi}{dt}$$

the induced e.m.f. is directly proportional to the rate of change of flux.

If there is no rate of change of flux, there is no induced e.m.f.

c.   *Establishes the relationships $E = Blv$ and $E = N \, d\phi/dt$ and used them to solve simple problems.*

From the above $E = N \dfrac{d\phi}{dt}$

Considering a conductor of length l and placed at right angle to a magnetic field as shown in Fig. 117. The conductor cuts the

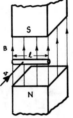

Fig. 117   A conductor cutting a magnetic field at right angles.

magnetic field with velocity v (m/s).

The distance moved in t seconds is given as vt. For a conductor $N = 1$

$$E = N \frac{d\phi}{dt} = N \frac{d(BA)}{dt}$$

$$E = 1 \frac{d(BA)}{dt}$$

since $A = l \, (vt) = l \, vt$

$$E = \frac{d}{dt} (B \, l \, v \, t)$$

since B, l and v are constants and t is the only variable, $Blv \dfrac{d}{dt} (t) = Blv$

$$E = Blv$$

E is the induced e.m.f. in the conductor, l, cutting a magnetic field of flux density B with a velocity v.

WORKED EXAMPLE 25

A conductor is moved at 15 m/s through a magnetic field having a flux density of 1 T.

Calculate (a) the e.m.f. induced per metre length of the conductor and (b) the force per metre length opposing this movement if the conductor is connected in a circuit having a total resistance of 1 Ω.

SOLUTION 25

(a) $E = Blv$

   $v = 15$ m/s, $B = 1$ T

   the e.m.f. incluced per metre, $\dfrac{E}{l} = Bv = 1 \times 15 = 15 \; \dfrac{\text{V}}{\text{m}}$.

(b) $F = BIl$

   $\dfrac{F}{l}$ = force per metre = $BI = 1 \times 15 = 15$ N/m

   where $I = \dfrac{V}{R} = \dfrac{15}{1} = 15$ A.

WORKED EXAMPLE 26

A coil of 1,000 turns is subjected to a flux change at the rate of 15 mWb/s. Determine the magnitude of the induced e.m.f.

SOLUTION 26

The magnitude of the induced e.m.f. is given by the formula

$$E = N \frac{d\phi}{dt}$$

$$E = 1000 \times 15 \times 10^{-3} = 15 \text{ volts}$$

where $\dfrac{d\phi}{dt} = 15 \times 10^{-3}$ Wb/s.

d.   *Explains the historical and technical significance of Fara day's and Lenz's Laws.*

Fig. 115 extablishes an induced e.m.f. such that it opposes the motion. Fig. 116 establishes an induced e.m.f. such that it opposes again the motion. Figures 115 and 116 show clearly Lenz's law and this is indicated algebraically by a negative sign

$$E = -N \frac{d\phi}{dt} \qquad \qquad \text{... (1)}$$

If L is the selt-inductance of the coil in Fig. 115 and Fig. 116, there is an induced e.m.f. in the coil due to the change of current.

$$E = -L \frac{di}{dt} \qquad \qquad \text{... (2)}$$

The induced e.m.f. is directly proportional to the rate of change of current. If there were no change in current, the induced e.m.f. would be zero.

Equations (1) and (2) are known as the Faraday's laws of electromagnetism. From equation (2), the self-inductance may be defined

$$\text{Self-inductance} = \frac{\text{induced e.m.f.}}{\text{rate of change of current}}$$

$$L = \frac{E}{\dfrac{di}{dt}}$$

neglecting the negative sign in this case as it only indicates direction of induced e.m.f. or current.

$$E = -N \frac{d\phi}{dt} \qquad \text{and } E = -L \frac{di}{dt} \quad \text{are the}$$

laws due to Faraday and the negative sign is due to Lenz's Law.

WORKED EXAMPLE 27

A centre-tapped coil of 1,000 turns is mounted on a magnetic core and has a self-inductance of 50 mH. Calculate:

(i)   the induced e.m.f. when the flux changes at the rate of 40 mWb/s

(ii)   the self-inductance of each half of the coil and

(iii)   the rate of current change in the coil.

SOLUTION 27

A centre-tapped coil of 1000 turns is shown in Fig. 118.

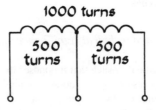

Fig. 118   Centre-tapped coil

(i) $E = -N \dfrac{d\phi}{dt} = -1000 \times 50 \times 10^{-3} = -50$ volts

(ii)   The total self-inductance = 50 mH, the self-inductance of each half coil is 25 mH.

(iii) $E = -L \dfrac{di}{dt}$

re-arranging the formula

$$\frac{di}{dt} = \frac{-50}{50 \times 10^{-3}} = -1,000 \text{ A/s.}$$

WORKED EXAMPLE 28

A current varying in the manner shown in the diagram of Fig. 119 is passed through a coil having an inductance of 500 mH.

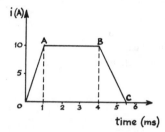

Fig. 119   Current/time graph

Calculate the e.m.f. induced in the coil during the periods OA, AB and BC.

SOLUTION 28

During the period OA

$$E = -L \frac{di}{dt} = -0.5 \times \frac{(10 - 0)}{(1 - 0) \times 10^{-3}} = -5,000 \text{ volts}$$

during the period AB

$$E = -0.5 \frac{(10 - 10)}{(4 - 1) \times 10^{-3}} = -0.5 \times \frac{0}{3 \times 10^{-3}} = 0 \text{ V}$$

during the period BC

$$E = -0.5 \frac{(0 - 10)}{(5.5 - 4) \times 10^{-3}} = \frac{5 \times 10^{3}}{1.5} = -3,333 \text{ volts.}$$

This example clearly establishes Lenz's Law.

e.   Describes the concept of eddy currents and eddy current loss, explaining their significance under conditions of a.c. magnetisation

Coils normally are wound on ferromagnetic materials in order to increase the self-inductance. If the ferro-magnetic materials are solid, eddy currents flow in the solid material which cause a power loss and thus decreases the efficiency of the device.

To reduce these "Eddy currents", the resistance of the core is increased by replacing it with thin painted laminations.

Eddy currents are objectionable in generator and motor armature and transformer cores, in which they cause power losses and hence the efficiencies are decreased.

Understands the concepts of self-and mutual inductance and relates these to the transformer principle.

f. *Defines self-inductance of a coil in terms of the proportionality of flux linkages and current in a linear magnetic medium and describes the production of induced voltage due to change in flux linkages.*

SELF INDUCTANCE OF COIL

$$E = -L \frac{di}{dt} = -N \frac{d\phi}{dt}$$

$$L = \frac{E}{\dfrac{di}{dt}} = \frac{\text{induced e.m.f.}}{\text{rate of change of current}}$$

L = self-inductance

at steady state

$$LI = N\phi$$

$$L = \frac{N\phi}{I}.$$

The self-inductance is directly proportional to the flux linkages ($N\phi$) and inversely proportional to the current.

WORKED EXAMPLE 29

The flux linked with a coil changes steadily from 5 mWb to 75 mWb in 7 ms. The average value of induced e.m.f. is 150 V. How many turns are there on the coil?

SOLUTION 29

$$E = N \frac{d\phi}{dt} = \text{the average induced e.m.f.}$$

$$150 = N \frac{(75 \times 10^{-3} - 5 \times 10^{-3})}{7 \times 10^{-3}} = N \frac{70}{7} = 10\,N$$

$N = 15$ the number of turns.

WORKED EXAMPLE 30

A current of 200 mA flowing in a 75 mH coil is reversed linearly in 150 ms. Calculate the value of the e.m.f. induced in the coil.

SOLUTION 30

$$E = -L \frac{di}{dt} = 75 \times 10^{-3} \frac{(-700 - 200)}{150 \times 10^{-3}} \times 10^{-3}$$

$$E = \frac{75 \times 400}{150} \times 10^{-3} = 0.2 \text{ volts.}$$

b. *Deduces and applies the relationship* $L = \dfrac{N^2}{S}$ *and* $E = L\dfrac{di}{dt}$

DESIGNING A COIL

$$L = \frac{N\phi}{I} \text{ at steady state, } H = \frac{IN}{l} \text{ and } S = \frac{l}{\mu_o \mu_r A}$$

$$L = \frac{NBA}{I} = \frac{NBAN}{Hl} = \frac{N^2 BA}{Hl} = N^2 \mu_o \mu_r \frac{A}{l} = \frac{N^2}{l/\mu_o \mu_r A} = \frac{N^2}{S}$$

the self-inductance depends greatly on the number

$$L = \frac{N^2}{S}$$

of turns and it is inversely proportional to the reluctance.

$$L = N^2 \mu_o \mu_r \frac{A}{l}$$

and since μ_o, A and l are constants then the self-inductance depends on N^2 and the type of material with relative permeability, μ_r. A coil wound on a ferro-magnetic material gives a much higher self-inductance than when it is wound on a non-magnetic material.

c. *Defines mutual inductance and describes the production of induced voltage due to change of mutual flux linkage.*

MUTUAL INDUCTANCE BETWEEN TWO COILS

Mutual inductance only exists between two coils. Consider two coils A and B which are mutually coupled as show in Fig. 120, where M is the mutual inductance.

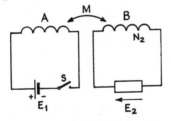

Fig. 120   Two coils closely coupled with mutual inductance M.

Two coils are said to have a mutual inductance of 1 henry if an e.m.f. of 1 volt is induced in the secondary when the primary current changes at the rate of 1 ampere per second.

When the switch S is closed, the current in coil A changes, the induced e.m.f. in coil B is given as

$$E_2 = M \frac{di_1}{dt} = N_2 \frac{d\phi_2}{dt} .$$

WORKED EXAMPLE 31

Two coils A and B are mutually coupled, the mutual inductance between them being 0.25 H. If the coils have 500 turns each, determine the induced e.m.f. in coil B when

(i)   the current in coil A changes at 75 A/s

(ii) the flux linking the coil changes at 0.5 Wb per second.

SOLUTION 31

(i) $E_2 = M \dfrac{di_1}{dt} = 0.25 \times 75 = 18.75$ volts

(ii) $E_2 = N \dfrac{d\phi_2}{dt} = 500 \times 0.5 = 250$ volts.

COILS CONNECTED IN SERIES AIDING

Two coils A and B are connected on an iron former as shown in Fig. 121 in series aiding

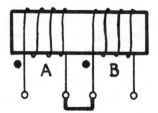

Fig. 121   Two coils connected in series aiding.

If the self-inductance of A is L_A and the self-inductance of B is L_B, the total self-inductance is given by

$$L = L_A + L_B + 2M$$

where M is the mutual inductance of the two coils.

COILS CONNECTED IN SERIES OPPOSING

Two coils A and B are connected on an iron former as shown in Fig. 122

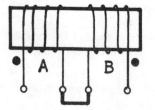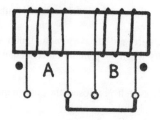

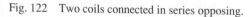

Fig. 122   Two coils connected in series opposing.

If again the self-inductance of A is L_A and the self-inductance of B is L_B, the total self-inductance is given by

$$L = L_A + L_B - 2M.$$

WORKED EXAMPLE 32

Two coils, *A* and *B*, have self-inductances of 10 mH and 20 mH respectively. When connected in series aiding the total inductance is 35 mH. Calculate the mutual inductance and estimate the total inductance of the coils when they are connected in series opposing.

SOLUTION 32

Referring to Fig. 121

$L_A = 10$ mH and $L_B = 20$ mH

$L = L_A + L_B + 2M$

$35 \times 10^{-3} = 10 \times 10^{-3} + 20 \times 10^{-3} + 2M$

$(35 - 30) \times 10^{-3} = 2M$

$M = 2.5$ mH

Referring to Fig. 122

$L = L_A + L_B - 2M$

$\quad = 10 \times 10^{-3} + 20 \times 10^{-3} - 2 \times 2.5 \times 10^{-3}$

$\quad = 30 \times 10^{-3} - 5 \times 10^{-3}$

$L = 25 \times 10^{-3}$ H

therefore

$L = 25$ mH

INDUCTIVE AND NON INDUCTIVE CIRCUITS

Straight copper wires possess a very small self-inductance, if the copper wires are in the form of a coil, the self-inductance is more predominant.

To increase the self-inductance the coil is wound on a magnetic circuit.

Therefore straight copper wires, coils on non-magnetic and coils on magnetic circuits are inductive.

Non-inductive circuits such as high stability resistors and incandescent lamps are made of wires bent back on itself since the field due to current flowing in one direction cancels that due to current flowing in opposite direction Fig. 123 shows the construction of a non-inductive resistor.

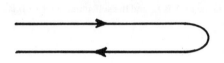

Fig. 123   Non-inductive resistor. Wire is bent back on itself.

d.   Describes the transformer principle in terms of Lenz's Law and induced volts per turn, deducing the effect of turns ratio on voltage ratio.

THE TRANSFORMER PRINCIPLE

Two coils of turns N_p and N_s are wound on a ferro-magnetic core as shown in Fig. 124.

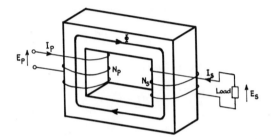

Fig. 124   The basic principle of a transformer.

The coil N_p is connected to an a.c. supply E_p and the coil N_s is connected to a load. The former coil is called *the primary* and the latter coil is called *the secondary*. Applying Faraday's laws of electromagnetism

$$E_p = -N_p \; \frac{d\phi}{dt} \; \dots (1) \qquad \text{and } E_s = -N_s \; \frac{d\phi}{dt} \; \dots (2)$$

assuming that the rate of change of flux is the same, that is, no flux leakage occurs dividing equation (1) by equation (2)

$$\frac{E_p}{E_s} = \frac{-N_p \dfrac{d\phi}{dt}}{-N_s \dfrac{d\phi}{dt}} = \frac{N_p}{N_s}$$

$$\frac{E_p}{E_s} = \frac{N_p}{N_s} \quad \dots (3)$$

Therefore, the ratio of the induced e.m.f. of the primary to the induced e.m.f. of the secondary is equal to the turns ratio.

Assuming no power loss between the primary and secondary, the primary volt-amperes are equal to the secondary volt-amperes

$$E_p I_p = E_s I_s$$

or $\dfrac{E_p}{E_s} = \dfrac{I_s}{I_p} \quad \dots (4)$

combining equations (3) and (4), we have

$$\frac{E_p}{E_s} = \frac{N_p}{N_s} = \frac{I_s}{I_p} \quad \dots (5)$$

e. *Deduces that energy stored in an inductor is $\frac{1}{2}LI^2$.*

Power is the rate of doing work

  P = power in watts

  W = energy or work in joules

  $P = \frac{W}{t}$ or $W = Pt = EIt = IRIt = I^2Rt$.

When an inductor is completely demagnetised and is connected across a d.c. supply, the current initially is zero and hence the energy is zero, t seconds later the inductor is energised and the current has grown to a finite value i, the induced e.m.f. is $L\frac{di}{dt}$ and its energy will be $\left(L\frac{di}{dt}\right)i\,dt$.

The energy stored in the inductor $= \int_0^I \left(L\frac{di}{dt}\right)i\,dt = \int_0^I Li\,di = \left[\frac{1}{2}Li^2\right]_0^I = \frac{1}{2}LI^2$.

The current grows from $i = o$, to $i = I$, the final steady current.

The energy stored in an inductor from $i = o$ to $i = I$ is $W = \frac{1}{2}LI^2$, the energy stored in an inductor at any instant is $W = \frac{1}{2}Li^2$ where i is the instantaneous current.

WORKED EXAMPLE 33

(i)   Calculate the e.m.f. induced in an inductance of 500 mH when the current is changing at a rate of 200 A/s.

(ii) Calculate the energy stored in this inductance when the current is 5 A.

SOLUTION 33

(i) $E = -L\frac{di}{dt} = -500 \times 10^{-3}\,200 = -50$ volts

(ii) $W = \frac{1}{2}Li^2 = \frac{1}{2} \times 500 \times 10^{-3} \times 5^2 = 6.25$ joules.

WORKED EXAMPLE 34

Calculate the energy contained in an inductance of 5 H carrying a current of 1 A.

SOLUTION 34

  $W = \frac{1}{2}LI^2 = \frac{1}{2}\,5 \times 1^2 = 2.5$ J.

f. *Solves problems on self-inductance, mutual inductance and the transformer principle.*

THE RESISTANCE VIEWED AT THE PRIMARY OF A TRANSFORMER

Consider a simple transformer in Fig. 125 whose a.c. input resistance is required

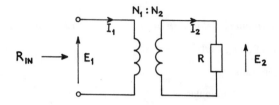

Fig. 125   The a.c. input resistance at the primary of the transformer

From the equation $\dfrac{E_1}{E_2} = \dfrac{N_1}{N_2} = \dfrac{I_2}{I_1}$ we have

$$E_1 = \left(\frac{N_1}{N_2}\right) E_2 \; ... \; (1) \text{ and } \qquad I_1 = I_2 \; \left(\frac{N_2}{N_1}\right) \; ... \; (2)$$

and load $R = \dfrac{E_2}{I_2} \; ... \; (3)$

The a.c. input resistance viewed at the primary terminals, R in is given

$$R_{in} = \frac{E_1}{I_1} = \frac{\text{Input voltage}}{\text{Input current}}$$

substituting in this equation, equations (1) and (2)

$$R_{in} = \frac{E_2 \left(\dfrac{N_1}{N_2}\right)}{I_2 \left(\dfrac{N_2}{N_1}\right)}$$

$$R_{in} = \frac{E_2}{I_2} \left(\frac{N_1}{N_2}\right)^2$$

and finally substituting $R = \dfrac{E_2}{I_2}$, we have

$$R_{in} = \left(\frac{N_1}{N_2}\right)^2 R \; ... \; (4)$$

the resistance viewed at the primary.

WORKED EXAMPLE 35

An ideal transformer having a 50:1 step down ratio, has a load current and voltage of 0.2 A and 50 V respectively. Calculate the primary current and voltage.

SOLUTION 35

Using equation

$$\frac{I_1}{I_2} = \frac{N_2}{N_1}$$

or $\quad I_1 = I_2 \dfrac{N_2}{N_1}$

$$I_1 = 0.2 \frac{1}{50} = 4 \text{ mA}$$

Using equation

$$\frac{E_1}{E_2} = \frac{N_1}{N_2}$$

or $\quad E_1 = \frac{N_1}{N_2} E_2 = 50 \times 50 = 2,500 \text{ V.}$

WORKED EXAMPLE 36

A 5:1 step down transformer has a primary voltage supply of 240 V and a secondary load resistance of 1,000 Ω. Calculate : (i) the resistance at the primary terminals

(ii) the secondary voltage and

(iii) the primary and secondary currents

SOLUTION 36

(i) $\quad R_{in} = \left(\frac{N_1}{N_2}\right)^2 R$

$\quad R_{in} = 5^2 \times 1,000 = 25,000 \ \Omega$

(ii) $\quad E_2 = E_1 \frac{N_2}{N_1} = 240 \times \frac{1}{5} = 48 \text{ volts}$

(iii) $I_2 = \frac{E_2}{R} = \frac{48}{1,000} = 48 \text{ mA}$

$\quad I_1 = \frac{E_1}{R_{in}} = \frac{240}{25,000} = 9.6 \text{ mA.}$

WORKED EXAMPLE 37

A coil is wound with 1,000 turns. A current of 2 amperes flowing in the coil produces a flux of 100 μWb. Determine the value of the self-inductance.

SOLUTION 37

At steady state

$$LI = N\phi \qquad \qquad \dots (1)$$

from the equations $E = -L \dfrac{di}{dt} = -N \dfrac{d\phi}{dt}$

from equation (1)

the self-inductance $L = \dfrac{N\phi}{I}$

$$L = \frac{1,000 \times 100 \times 10^{-6}}{2}$$

$$L = 50 \text{ mH.}$$

WORKED EXAMPLE 38

A coil having an inductance of 225 mH is carrying a current of 20 A. Calculate the induced e.m.f. in the coil when the current is:-

(i)   reduced to zero in 0.05 seconds

(ii) reversed in 0.05 seconds

(iii) increased to 50 A in 3 seconds.

SOLUTION 38

(i) $\quad E = -L\ \dfrac{di}{dt} = -225 \times 10^{-3}\ \dfrac{(0-20)}{0.05} = 90$ V

(ii) $\quad E = -L\ \dfrac{di}{dt} = -225 \times 10^{-3}\ \dfrac{(-20-20)}{0.05} = 180$ V

(iii) $\quad E = -L\ \dfrac{di}{dt} = -225 \times 10^{-3}\ \dfrac{(50-20)}{3} = 2.25$ V.

SUMMARY 5

Left and Right Hand Rules are shown in Fig. 126 and Fig. 127.

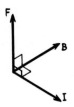

Fig. 126   Left Hand Rule

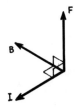

Fig. 127   Right Hand Rule

1. $F = BIl \sin \theta$

2. $E = Blv$

3. $E = -L\ \dfrac{di}{dt}$

4. $E = -N\ \dfrac{d\phi}{dt}$

5. $LI = N\phi$ steady stale

6. $E_2 = M\ \dfrac{di}{dt}$

7. $E_2 = N_2\ \dfrac{d\phi_2}{dt}$

8. $W = \dfrac{1}{2} LI^2$

9. $\dfrac{E_p}{E_s} = \dfrac{N_p}{N_s}$

10. $\dfrac{E_p}{E_s} = \dfrac{I_s}{I_p}$

F = force on a conductor (N)

B = flux density (T)

I = Current (A)

l = length of conductor (l)

θ = angle between the straight conductor and magnetic field (degrees)

E = induced e.m.f. (V)

v = velocity (m/s)

L = self-inductance (H)

$\dfrac{di}{dt}$ = rate of change of current (A/s)

$\dfrac{d\phi}{dt}$ = rate of change of flux (Wb/s)

N = number of turns

M = mutual inductance (H)

11. $R_{in} = \left(\dfrac{N_1}{N_2}\right)^2 R_{load}$

W = energy stored (J)
E_p = primary voltage (V)
E_s = secondary voltage (V)
R_{in} = a.c. resistance at the input terminals of the transformer (Ω)
$\dfrac{N_1}{N_2}$ = turns-ratio
R_{load} = load resistance (Ω).

EXERCISE 5

1. The force on a conductor of 0.5 m is 0.25 N and the conductor carries a current of 50 mA and is placed at right angles to a magnetic field. Determine the flux density of the magnetic field. (Ans. 10 T).

2. A conductor 40 cm in length and carrying a current of 150 mA is placed at 80° to a magnetic field having a flux density of 0.9 T. What is the angle of the conductor to the magnetic field having the same flux density and the force on the conductor is 1/5 that of the first case. (Ans. 11.32°).

3. A conductor at right angles to a uniform magnetic field of flux density, 50 mT has 10 cm of its length in the field. Calculate the current required to cause a force of 100 mN to act on the conductor. (Ans. 20 A).

4. A rectangular coil 20 cm long and 15 cm wide and consisting of one turn is rotated about its longer axis at 500 r.p.m. in a uniform magnetic field of 1 T field is situated normal to the axis. Determine the e.m.f. induced in the single turn coil, and in a similar coil of 100 turns. (Ans. 0.79 V, 79 V).

5. An e.m.f. of 5 volts is induced in a coil due to a flux change at the rate of 0.5 Wb/s. Calculate the number or turns of the coil. (Ans. 10).

6. Calculate the e.m.f. induced in an inductance of 200 mH when the current is changing at a rate of 500 A/s. (Ans. 100).

7. With the aid of two coils closely coupled together explain what is meant by mutual inductance. Two coils A and B are mutually coupled, the mutual inductance between them being 0.2 H. Calculate the e.m.f. in coil B when the current in coil A changes at 100 A/s. (Ans. 20 V).

8. An ideal step up transformer having a turns ratio of 10:1 is connected to a supply voltage of 240 V. The secondary load is a resistor of 12 KΩ. Calculate:

   (i) the secondary voltage and current

   (ii) the primary current

   (iii) the resistance viewed at the primary terminals. (Ans. (i) 2,400 V, 0.2 A (ii) 2 A, (iii) 120 Ω).

9 (i) Show that the self-inductance of a coil is directly proportional to the number of turns squared and the permeability of the magnetic material.

   (ii) Show that the reluctance of non-magnetic material is constant for all values of current.

   (iii) Show that the reluctance of a magnetic material varies with the current.

10. Explain briefly and clearly the construction of inductive and non-inductive components using conducting wires.

ALTERNATING VOLTAGES AND CURRENTS

6 . *Displays waveforms and determines the main parameters used to describe and measure them.*

    a. *Defines the terms amplitude, period, frequency, instantaneous, peak-to-peak r.m.s average in relation to alternating (sinusoidal and non-simusoidal) and unidirectional waveforms and used an oscilloscope to display and measure these parameters.*

ALTERNATING SINUSOIDAL WAVEFORMS

Fig. 128 shows an alternating sinusoidal waveform of two cycles, the y-axis is $A \sin \omega t$ and the x-axis is ωt.

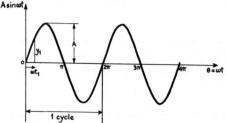

Fig 128   A.C. sinusoidal waveform.

Trigonometrically this waveform is represented by the equation

    $y = A \sin \omega t$

The waveform commences at the origin O.

AMPLITUDE

The maximum or peak value is called the amplitude and it is represented by A.

PERIOD

The period is the time taken to trace out one complete cycle and it is represented by T.

FREQUENCY

The frequency is the number of cycles traced out in one second and is represented by f.

INSTANTANEOUS

The instantaneous value of y is that value at a certain instant t, (t_1 for example).
At t_1, $y = y_1 = A \sin \omega t_1$, y_1 is the instantaneous value of y at $t = t_1$.

PEAK TO PEAK

The peak-to-peak value is the value between the maximum or peak positive value and the maximum of peak negative value. In this case, the peak-to-peak value 2 A.

R.M.S. (ROOT MEAN SQUARE)

The r.m.s. is the square root value of the mean squares. This is best illustrated as follows:

$\theta°$	0	30	60	90	120	150	180	210	240	270	300	330	360
$A \sin \theta°$	0	$\frac{1}{2}A$	$\frac{\sqrt{3}}{2}A$	A	$\frac{\sqrt{3}}{2}A$	$\frac{1}{2}A$	0	$-\frac{1}{2}A$	$-\frac{\sqrt{3}}{2}A$	$-A$	$-\frac{\sqrt{3}}{2}A$	$-\frac{1}{2}A$	0
$A^2 \sin^2 \theta°$	0	$\frac{1}{4}A^2$	$\frac{3}{4}A^2$	A^2	$\frac{3}{4}A^2$	$\frac{1}{4}A^2$	0	$\frac{1}{4}A^2$	$\frac{3}{4}A^2$	A^2	$\frac{3}{4}A^2$	$\frac{1}{4}A^2$	0

The means squares $= \sqrt{\dfrac{0^2 + \frac{1}{4}A^2 + \frac{3}{4}A^2 + A^2 + \frac{3}{4}A^2 + \frac{1}{4}A^2 + 0 + \frac{1}{4}A^2 + \frac{3}{4}A^2 + A^2 + \frac{3}{4}A^2 + \frac{1}{4}A^2 + 0}{12}}$

$= \sqrt{\dfrac{6A^2}{12}} = \dfrac{A}{\sqrt{2}} = 0.707$ A approximately.

AVERAGE

The average value of one complete cycle is zero since the positive area of the half cycle cancels the negative area of the other half cycle.

The average value of one half cycle is given as $\dfrac{2A}{\pi} = 0.637$ A approximately.

NON SINUSOIDAL WAVEFORMS

Non sinusoidal waveforms such as the triangular and square waveforms are shown in Fig. 129 and Fig. 130 respectively.

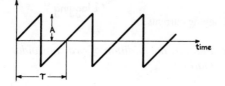

Fig. 129   Triangular waveforms.

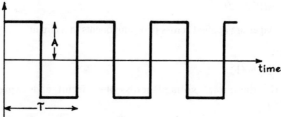

Fig. 130   Square waveforms.

These waveforms are called complex or non-sinusoidal waves. The amplitude and the period are shown on the diagrams.

b.   *Defines form factor and determines the approximate average and r.m.s. value of given sinusoidal and non-sinusoidal waveforms.*

A sinusoidal waveform

Form factor $= \dfrac{\text{r.m.s. value}}{\text{average value}} = \dfrac{0.707 \text{ A}}{0.637 \text{ A}}$

$= 1.11.$

For any sinusoidal current or voltage the form factor is the same and it is always 1.11 approximately to three significant figures.

For a non-sinusoidal waveform the form factor is different and should be determined for each waveform.

Uses phasor and algebraic representation of sinusoidal quantities.
a. Defines a phasor quantity.

PHASOR OR PHASOR DIAGRAM

A phasor quantity has magnitude and direction.

A current in a circuit is 3 A and leads the voltage by 25°. If the voltage is taken as a reference, a horizontal line of indefinite magnitude, then three units are scaled along a positive direction of 25° to the voltage. If the current lags by 25°, then three units are scaled along a negative direction of 25°. Fig 131 represents the phasor quantity of current with respect to the voltage.

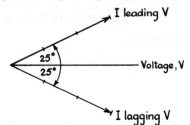

Fig. 131   Leading and Lagging currents.

b. Determines the resultant or the addition of two sinusoidal voltages by graphical and phasor representation.

PHASOR REPRESENTATION

WORKED EXAMPLE 39

Two sinusoidal voltages are applied across two components in a circuit

$E_1 = 10 \sin \omega t$
$E_2 = 20 \sin (\omega t + \pi/2)$.

Determine the resultant total voltage across the circuit and express in the form $E_R = E_m \sin (\omega t \pm \phi)$.

SOLUTION 39

Since E_1 is a reference phasor quantity, it is drawn 10 units horizontally and since E_2 is 90° leading or $\pi/2$ leading, it is drawn up perpendicularly 20 units as shown in Fig. 132.

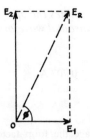

Fig. 132   Phasor representation of voltages E_1 and E_2 and their resultant E_R.

The resultant in this case is the hypotenuse of the right angled triangle.

$$E_m^2 = E_{1m}^2 + E_{2m}^2$$

$$E_m = \sqrt{20^2 + 10^2} = 22.36.$$

From the right - angled triangle OE_1E_R $\tan\theta = \dfrac{20}{10} = 2$ and hence $\theta = 63.43°$.

The resultant is given by

$$E_R = 22.36 \sin \quad (\omega t + 63.43°)$$

but ω is the angular velocity in radians per second, ωt is the angle in radian and radian and it is better to express θ in radians

$$63.43° = 63.43° \times \frac{\pi}{180°} = 1.11^c$$

$$E_R = 22.4 \sin \quad (\omega t + 1.11^c)$$

the resultant voltage in three significant figures.

GRAPHICAL REPRESENTATION

WORKED EXAMPLE 40

If $E_1 = 10 \sin \omega t$ and $E_2 = 20 \sin (\omega t + \pi/2)$ determine graphically $E_1 + E_2$.

SOLUTION 40

θ^c	0	$\frac{\pi}{6}$	$\frac{\pi}{3}$	$\frac{\pi}{2}$	$2\frac{\pi}{3}$	$5\frac{\pi}{6}$	π	$7\frac{\pi}{6}$	$4\frac{\pi}{3}$	$3\frac{\pi}{2}$	$5\frac{\pi}{3}$	$11\frac{\pi}{6}$	2π
$\sin\theta^c$	0	0.5	0.866	1	0.866	0.5	0	−0.5	−0.866	−1	−0.866	−0.5	0
$E_1 = 10\sin\theta^c$	0	5	8.66	10	8.66	5	0	−5	−8.66	−10	−8.66	−5	0
$\sin\left(\theta^c + \frac{\pi}{2}\right)$	1	0.800	0.5	0	−0.5	−0.866	−1	−0.866	−0.5	0	0.5	0.866	1
$E_2 = 20\sin\left(\theta^c + \frac{\pi}{2}\right)$	20	17.32	10	0	−10	−17.32	−20	−17.32	−10	0	+10	17.32	20
$E_1 + E_2$	20	22.3	18.7	10	−1.34	−12.3	−20	−22.3	−18.7	−10	−1.34	12.3	20

Fig. 133 shows the plot of the resultant waveform, $E_1 + E_2$ against ωt.

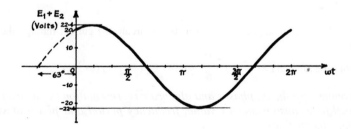

Fig. 133   Resultant waveform for E_1 and E_2, leading by 63°.

The amplitude of the resultant waveform is approximately 22.4 and the waveform lead by about 1.1^c.

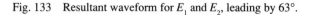

c.   Explains the phase-angle relationship between two alternating quantities.

PHASE ANGLE LEADING AND LAGGING

Fig. 134 shows a lagging, *C*, and a leading, *B*, waveforms with respect to the references, *A*, waveform.

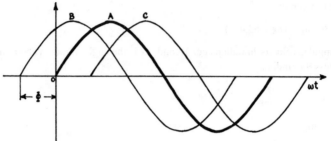

Fig. 134   Reference, leading and lagging waveforms.

A is taken as a reference waveform since it commences at the origin.

B is leading *A* since the peak value of the waveform is reached before that of *A*. *C* is lagging *A* since the peak value of the waveform is reached after that of *A*.

d.   Defines a sinusoidal voltage in the form $v = V_m \sin(\omega t + \phi)$.

The instantaneous value of the voltage is given in terms of the maximum value, the angular velocity, time and phase angle.

v $= V_m \sin(\omega t \pm \phi)$ volts
v = instantaneous value of the voltage at time t
V_m = the peak or maximum value of voltage
ω = the angular velocity in radians per second
t = time in seconds
ϕ = phase angle in radians.

When the sign in front of ϕ is positive then the waveform leads the reference, when the sign in front of ϕ is negative then the waveform lags the reference.

e.   Determines current from the application of sinusoidal voltage to a resistive circuit.

If a sinusoidal voltage is applied across a resistive circuit, the current through the circuit is sinusoidal.

If $v = V_m \sin(\omega t \pm \phi)$ then $i = \dfrac{v}{R}$ or $i = \dfrac{V_m}{R} \sin(\omega t \pm \phi) = I_m \sin(\omega t \pm \phi)$.

f.   Interrelates graphical, phasor and algebraic representation in the determination of amplitude, instantaneous value, frequency period and phase of sinusoidal voltage and currents.

WORKED EXAMPLE 41

An alternating current is given by the expression $i = 10 \sin(3142t + \dfrac{\pi}{4})$ A.

Calculate: (i)   the peak value
             (ii) the average value

    (iii) the r.m.s. value

    (iv) the frequency

    (v) the periodic time

    (vi) the angular velocity

    (vii) the instantaneous value of the current when $t = 0.25$ ms.

SOLUTION 41

(i) $I_{max} = 10$ A

(ii) $I_{av} = 10 \times \dfrac{2}{\pi} = 6.37$ A

(iii) $I = \dfrac{10}{\sqrt{2}} = 7.07$ A

(iv) $f = \dfrac{\omega}{2\pi} = \dfrac{3142}{2\pi} = 500$ Hz

(v) $T = \dfrac{1}{f} = \dfrac{1}{500} = 2$ ms

(vi) $\omega = 3{,}142$ radians per second

(vii) $i = 10 \sin (3{,}142 \times 0.25 \times 10^{-3} + \dfrac{\pi}{4})$

      $= 10 \sin (0.7854^{c} + 0.7854^{c})$

      $= 10 \sin (1.5708^{c})$

      $= 10$ A.

WORKED EXAMPLE 42

An instantaneous voltage is given by the expression $v = 100 \sin (3{,}142t + \dfrac{\pi}{6})$.

(a) Calculate: (i) the frequency

                   (ii) the period

                   (iii) the r.m.s. value

                   (iv) the average value

                   (v) the form factor

                   (vi) the instantaneous value after 1 ms.

(b) Represent the above voltage and the instantaneous current $i = 100 \sin 3{,}142t$ A in a phasor diagram.

SOLUTION 42

(a) (i) $f = \dfrac{\omega}{2\pi} = \dfrac{3{,}142}{2\pi} = 500$ Hz

     (ii) $T = \dfrac{1}{f} = \dfrac{1}{500} = 2$ ms

     (iii) $V_{r.m.s.} = \dfrac{V_{max}}{\sqrt{2}} = \dfrac{100}{\sqrt{2}} = 70.7$ volts

     (iv) $V_{av} = \dfrac{2}{\pi} V_{max} = 2 \times \dfrac{100}{\pi} = 63.7$ volts

(v) Form factor $= \dfrac{\text{r.m.s.}}{\text{average}} = \dfrac{70.7}{63.7} = 1.11$

(vi) $v = 100 \sin (3{,}142 \times 1 \times 10^{-3} + \dfrac{\pi}{6}) = 100 \sin \dfrac{7\pi}{6}$

$= 100 \left(-\dfrac{1}{2}\right) = -50$ volts.

(b) $i = 100 \sin 3{,}142 \, t$ A

this is taken as a reference since we draw a horizontal phasor of 100 V equivalent to 5 cm (that is 20 V per cm is the scale). The voltage is drawn at 30° to the horizontal and equal to 5 cm.

Fig. 135 represents the current and voltage in a phasor diagram.

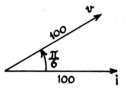

Fig. 135 v leads i by 30° or $\dfrac{\pi}{6}$.

WORKED EXAMPLE 43

(a) Write down the mathematical expression representing a sinusoidal current having an r.m.s. value of 250 mA and frequency 2,500 Hz and leading a voltage by a phase angle of 75°.

(b) Write down the mathematical expression representing a sinusoidal voltage having a peak value of 100 V and a periodic time of 2.5 ms and lagging a current by a phase angle of $\pi/9$.

SOLUTION 43

(a) $i = 250 \times 10^{-3} \sqrt{2} \sin (2\pi \, 2{,}500 \, t + 5 \dfrac{\pi}{12})$ A.

where $75° = 75° \times \dfrac{\pi}{180°} = \dfrac{5\pi^c}{12}$

$i = \sin (15{,}708t + 5 \dfrac{\pi}{12})$ A

(b) $v = 100 \sin (2\pi \dfrac{1{,}000}{2.5} t - \dfrac{\pi}{9})$ V

where $T = \dfrac{1}{f} = 2.5$ ms

$f = \dfrac{1}{2.5 \times 10^{-3}} = \dfrac{1000}{2.5} = 400$ Hz

$v = 100 \sin (2513.3t - \dfrac{\pi}{9})$.

g.   *Determines power in an a.c. resistive circuit from given data.*

Power in a resistive a.c. circuit is given by the r.m.s. value of current squared times the resistance of the resistor

$P = I^2 R$

where I = r.m.s. value of current

or $\quad P = \dfrac{V^2}{R}$

where V = r.m.s. value of voltage.

The power absorbed by a resistive load in an a.c. circuit is given simply by the product of the r.m.s. values of voltage and current.

$P = IV$

where I and V are the r.m.s. values of current and voltage.

Since the phase difference between the voltage and current waveforms in an a.c. resistive circuit is zero then cos 0 = 1 and $P = IV$. But the phase difference between I and V in non-resistive components is finite and this affects the power absorbed by the load. This will be covered in dedail in the next chapter.

If $v = V_m \sin \omega t$ is applied to a circuit of resistance R Ω, the resulting current is represented by $i = I_m \sin \omega t$.

The power at any instant $= P = IV = I_m \sin \omega t\; V_m \sin \omega t$

Mean power over a cycle $\quad = $ (mean value of i^2) $\times R$

$\quad = $ (mean value of $I_m^{\,2} \sin^2 \omega t$) $\times R$

$$= \frac{I_m^{\,2}}{2} R = \frac{I_m}{\sqrt{2}} \times \frac{I_m}{\sqrt{2}} \times R$$

$P = I \times I \times R = I^2 R = VI.$

Fig. 136 Shows the instantaneous values of current, voltage and power

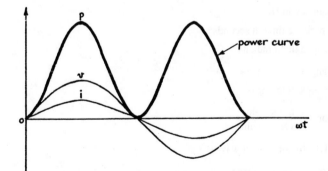

Fig. 136　Power curve for an a.c. resistive curve.

Equation of power curve is given by

$p = iv = I_m V_m \sin^2 \omega t$

$$= \frac{I_m V_m}{2} (1 - \cos 2\omega t).$$

WORKED EXAMPLE 44

A current wave of equation $i = 10 \sin \omega t$ mA is flowing through a 1,000 Ω non-inductive resistor. Calculate the p.d. across the circuit and the instantaneous value of the power at $\omega t = \pi/2$.

SOLUTION 44

The p.d. across the 1 KΩ resistor $v = i \times 1,000 = 10,000 \sin \omega t$ mV

$$= 10 \sin \omega t \text{ V}$$

The instantaneous value of the power $= p = iv$

$$p = 10 \times 10^{-3} \sin \omega t \times 10 \sin \omega t$$

$$= 0.1 \sin^2 \omega t$$

when $\omega t = \dfrac{\pi}{2}$, $\sin^2 \omega t = \sin^2 \dfrac{\pi}{2} = 1$ and $p = 0.1$ W.

SUMMARY 6

$$i = I_m \sin (\omega t \pm \phi)$$
$$v = V_m \sin (\omega t \pm \phi)$$

i, v are instantaneous values of current and voltage respectively

I_m, V_m are the peak values of current and voltage respectively

I, V are the r.m.s. values of current and voltage respectively

$$I = \frac{I_m}{\sqrt{2}} \qquad \text{and } V = \frac{V_m}{\sqrt{2}}$$

$\omega = 2\pi f$ = angular velocity radians per second

$\quad f$ = frequency in Hz

$\quad T = 1/f$ periodic time in seconds

$\quad \phi$ = phase angle in radians.

Positive phase angle (leading)

Negative phase angle (lagging)

$$\text{Form factor} = \frac{\text{r.m.s. value}}{\text{average value}}.$$

Form factor = 1.11 for sinusoidal waves.

Average value of current or voltage

$$I_{av} = \frac{2}{\pi} I_m \qquad\qquad V_{av} = \frac{2}{\pi} V_m.$$

$P = IV = I^2 R = V^2/R$ Power in a resistor.

EXERCISES 6

1. (i) Sketch one cycle of a non-sinusoidal waveform having a peak value of 50 volts and a period of 5 ms.

   (ii) Sketch one cycle of a sinusoidal current waveform having a 100 mA r.m.s. value and frequency of 1,000 Hz.

   (iii) Sketch one cycle of a sinusoidal voltage waveform having a peak value of 100 volts, a frequency of 50 Hz and a phase lag of 30°.

2. Draw phasors to represent the following voltages $v_1 = 15 \sin \omega t$, $v_2 = 25 \sin (\omega t - \pi/4)$. Determine by calculation the resultant voltage $v_1 + v_2$, in the form $v_R = V_{max} \sin (\omega t \pm \phi)$.

   (Ans. 37.15, 0.496° $V_R = 37.15 \sin (\omega t - 0.496°)$.

3. A sinusoidal alternating voltage has the trigonometric equation

   $v = 100 \sqrt{2} \sin (377\, t + \pi/4)$ V

Determine: (i) the peak value

   (ii) the average value

   (iii) the r.m.s. value

   (iv) the angular velocity

   (v) the frequency

   (vi) the periodic time

   (vii) the instantaneous voltage when t = 5 ms

   (viii) the phase angle, state that lags or leads.

   (Ans. (i) 141.4 V (ii) 90 V (iii) 100 V (iv) 377 rad/s (v) 60 Hz (vi) 16.7 ms (vii) 64.2 V (viii) $\pi/4$ leading.

4. An alternating voltage, $v = 10 \sin (500\, \pi t - \pi/6)$ is applied across a 100 Ω resistor, write down an expression for the instantaneous current. Sketch one cycle of the current and voltage on the same time base. (Ans. $i = 100 \sin (500\, \pi\, t - \pi/6)$ mA).

5. An instantaneous current in a circuit is given by the expression

   $i = 100 \sin (314,200\, t - \pi/12)$ mA.

   (a) Calculate : (i) its frequency

   (ii) its r.m.s. value

   (iii) its instantaneous value when $t = 2$ μs.

   (b) Calculate the power developed in 2.2 KΩ resistor when the above current flows through it.

   (c) If the instantaneous value of voltage supplied to the circuit is $v = 100 \sin 314,200\, t$ V draw a phasor diagram showing the r.m.s. values of voltage and current.

   (Ans. (a) (i) 50 KHZ (ii) 70.7 mA (iii) 35.8 mA (b) 11 W (c) I lags V by $\pi/12$).

6. By means of a phasor representation to a suitable scale, add the following voltages:

   $E_1 = 100 \sin \omega t$, $E_2 = 75 \sin (\omega t + \pi/3)$, $E_3 = 60 \sin (\omega t\, \pi/2)$.

Represent the resultant voltage in similar form.

(Ans. $E_R = 135 \sin(\omega t + 0.052c)$.

7. A sinusoidal voltage has an r.m.s. value of $100\sqrt{2}$ V at a frequency of 50 Hz. Commencing from zero in the cycle, and rising positively, find the time taken for the voltage to reach a value of 100 V. (Ans. 1.67 ms).

8. An instantaneous value of current is given as i = 200 $\sqrt{2}$ sin 877 t mA. Calculate:

    (i)   the r.m.s. value of the current

    (ii) the frequency

    (iii) the instantaneous value of the current when t is 0.001 second.

    (Ans.   (i) 0.2 A   (ii) 139.6 HZ   (iii) 0.22 A).

9. A non-inductive resistor takes $10\sqrt{2}$ A from a single phase 220 V mains, the voltage and current being sinusoidal. Plot graphs of currents voltage and power over one complete cycle. Determine the average power dissipated (i) by calculation and (ii) from the waveform of power. (Ans 3110.8 W).

10. Write down the mathematical expression representing a sinusoidal voltage having a peak value of 500 V and a periodic time of 10 ms and leading a current by a phase angle of $\pi/10$. (Ans. 500 (sin π t + $\pi/10$)).

SINGLE PHASE A.C. CIRCUITS

7. *Uses a.c. circuit theory to solve simple series a.c. circuit problems.*
    a. *Draws the phasor diagrams and related voltage and current waveforms for circuits comprising.*

- *Pure resistance*
- *Pure inductance*
- *Pure capacitance*

PURE RESISTANCE

An a.c. voltage source is applied across a pure resistor R. where V is the r.m.s. voltage and $I = V/R$ r.m.s. current, Fig. 137.

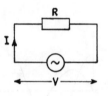

Fig. 137   Pure non-inductive resistance.

WAVEFORMS

Fig. 138 shows the current in phase with the voltage.

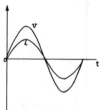

Fig. 138   In phase waveforms

PHASORS

Fig. 139 shows the phasor diagram. If I is the reference phasor then V is in phase with I.

Fig. 139   Phasor diagram for a pure resistance.

PURE INDUCTANCE

An a.c. voltage source is applied across a pure inductance where V is the r.m.s. voltage and I is the r.m.s. current. Taking I as the reference phasor, the voltage leads the current by 90°. Fig. 140 shows the circuit and Fig. 141 the relevant waveforms. The phasor diagram is shown in Fig. 142.

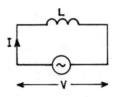

Fig. 140   Pure inductance.

WAVEFORMS

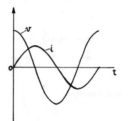

Fig. 141   Out of phase waveforms, i lags v.

PHASORS

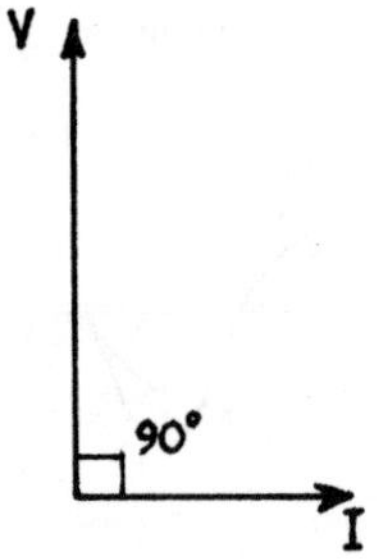

Fig. 142   Phasor diagram for a pure inductance.

PURE CAPACITANCE

An a.c. voltage source is applied across a pure capacitor where V is the r.m.s. voltage and I is the r.m.s. current, Fig. 143 shows the circuit. Taking I as the reference phasor, the voltage lags the current by 90°.

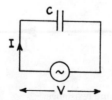

Fig. 143   Pure capacitance.

The waveforms and phasors are shown in Fig. 144 and Fig. 145.

WAVEFORMS

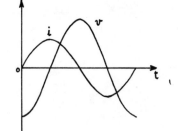

Fig. 144   Out of phase waveforms, I leads v

PHASORS

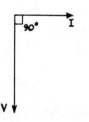

Fig. 145   Phasor diagram for a pure capacitance.

b.   *Describes inductive reactance and capacitive reactance in terms of impeding the flow of an alternating current and uses basic relationships to solve simple problems.*

INDUCTIVE REACTANCE

For Fig. 140

$\quad X_L$ = inductive reactance = $\dfrac{V}{I}$ (ohm's law)

$\quad X_L = 2\pi fL$ (Ω)

CAPACITIVE REACTANCE

For Fig. 143

$\quad X_c$ = capacitive reactance = $\dfrac{V}{I}$ (ohm's law)

$\quad X_c = \dfrac{1}{2\pi fC}$ (Ω).

WORKED EXAMPLE 45

(a)   A pure resistor is subjected to an a.c. source of 100 V r.m.s. and the current drawn is 10 A r.m.s. Calculate the resistance.

(b)   A pure inductor is subjected to an a.c. source of 100 V r.m.s. (f = 50 Hz) and the current is 10 A r.m.s. Calculate the reactance and hence the inductance.

(c)   A pure capacitor is subjected to an a.c. source of 100 V. r.m.s. (f = 50 Hz) and the current is 10 A r.m.s. Calculate the reactance and hence the capacitance.

SOLUTION 45

(a) $R = \dfrac{V}{I} = \dfrac{100}{10} = 10\ \Omega$

(b) $X_L = \dfrac{V}{I} = \dfrac{100}{10} = 10\ \Omega$

   $X_L = 2\ \pi f L$

   $L = \dfrac{X_L}{2\ \pi f} = \dfrac{10}{2\pi\ 50} = 31.8\ \text{mH}$

(c) $X_c = \dfrac{V}{I} = \dfrac{100}{10} = 10\ \Omega$

   $X_c = \dfrac{1}{2\ \pi f C}$

   $C = \dfrac{1}{2\ \pi f X_c} = \dfrac{1}{2\ \pi\ 50(10)} = 318.3\ \mu\text{F}.$

WORKED EXAMPLE 46

(a)   An inductor of 50 mH when subjected to an a.c. supply of 50 Hz draws 100 mA current. Calculate the inductive reactance and the a.c. voltage.

(b)   A capacitor of 500 μF when subjected to an a.c. supply of 150 Hz draws 200 mA current. Calculate the capacitive reactance and the a.c. voltage.

SOLUTION 46

(a) $X_L = 2\ \pi f L = 2\ \pi \times 50 \times 50 \times 10^{-3} = 15.7\ \Omega$

   $V = IX_L = 100 \times 10^{-3} \times 15.7 = 1.57\ \text{V}$

(b) $X_c = \dfrac{1}{2\ \pi f c} = \dfrac{1}{2\ \pi \times 150 \times 500 \times 10^{-6}} = 2.12\ \Omega$

   $V = IX_c = 200 \times 10^{-3} \times 2.12 = 0.424\ \text{V}.$

c.   *Draws phasor diagrams corresponding to L-R and C-R series circuits.*

L-R SERIES CIRCUIT

An inductor normally possesses an a.c. resistance R which is shown in this case as a resistor in series with a pure inductor. Fig. 146 shows the equivalent circuit of an inductor.

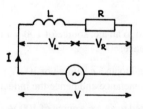

Fig. 146   L-R circuit.

THE PHASOR

If I in the reference vector, V_R is in phase with I and V_L is leading I by 90°. Fig. 147 shows the phasor of an inductor.

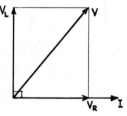

Fig. 147   Phasor diagram of an inductor.

The resultant vector is V as shown in Fig. 147. The voltage triangle is shown in Fig. 148.

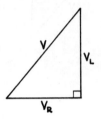

Fig. 148   Voltage triangle.

and applying Pythagoras theorem

$$V^2 = V_L^2 = V_R^2$$

$V_L = IX_L$ and $V_R = IR$

$$V^2 = I^2X_L^2 + I^2R^2 = I^2 \ (X_L^2 + R^2)$$
$$V = I \sqrt{X_L^2 + R^2}.$$

CR SERIES CIRCUIT

RC Circuit is shown in Fig. 149

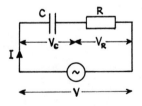

Fig. 149   RC circuit.

Fig. 149 shows the r.m.s. current I flowing through R and C. I is the reference vector since it is common through R and through C.

THE PHASOR

Fig. 150 shows the phasor diagram for the RC circuit in series. V_R is in phase with I and V_c lags I by 90°.

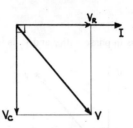

Fig. 150   The phasor diagram for a capacitor in series with a resistor.

The voltage triangle is shown in Fig. 151

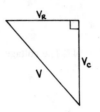

Fig. 151   Voltage triangle.

$$V^2 = V_R^2 + V_C^2$$

$V_C = IX_C$ and $V_R = IR$

$$V^2 = I^2R^2 + I^2X_C^2$$
$$V^2 = I^2 (R^2 + X_C^2)$$
$$V = I \sqrt{R^2 + X_C^2}.$$

d.   Defines impedance as Z = V/I.

IMPEDANCE

For Fig. 146 and Fig 149, impedance is defined as

$$Z = \frac{V}{I}.$$

For Fig. 146 $Z = \dfrac{V}{I} = \dfrac{I \sqrt{X_L^2 + R^2}}{I}$

$$Z = \sqrt{R^2 + X_L^2}$$

For Fig. 149 $Z = \dfrac{V}{I} = \dfrac{I \sqrt{R^2 + X_C^2}}{I}$

$$Z = \sqrt{R^2 + X_C^2}.$$

e.   Derives impedance triangle from voltage triangles and shows that $Z^2 = R^2 + X^2$ and that tan ϕ = X/R, sin ϕ = X/Z and cos ϕ = R/Z.

The impedance triangles are simply derived from the voltage triangle by dividing each side of the Fig. 148 and Fig. 151 by I since it is common.

Then we have Fig. 152 and Fig. 153 which are the impedance triangles.

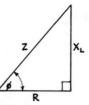

Fig. 152   Impedance triangle for *LR* circuit.

Fig. 153   Impedance triangle for *RC* circuit.

From the impedance triangles we get the following equations, by applying Pythagoras theorem.

$Z^2 = R^2 + X_L^2$ and $Z^2 = R^2 + X_C^2$.

If ϕ is the phase angle, then we have from each triangle of Fig. 152 and Fig. 153

$\tan \phi = X_L/R$	$\tan \phi = X_C/R$
$\cos \phi = R/Z$	$\cos \phi = R/Z$
$\sin \phi = X_L/Z$	$\sin \phi = X_C/Z$.

f.   Applies equations in c, d, and e to the solution of single branch L-R and C-R series circuits at power and radio frequencies.

WORKED EXAMPLE 47

In an R-C series circuit the a.c. p.ds across R and C are respectively 30 V and 40 V. Calculate the total a.c. p.d. applied.

SOLUTION 47

From the voltage triangle of Fig. 151
$$V^2 = V_R^2 + V_C^2$$
$$V = \sqrt{V_R^2 + V_C^2}$$
$$V_R = 30 \text{ V and } V_C = 40 \text{ V}$$
$$V = \sqrt{30^2 + 40^2} = \sqrt{900 + 1600}$$
$$= \sqrt{2,500} = 50 \text{ V}$$

WORKED EXAMPLE 48

In an L-R series circuit the a.c. supply voltage is 240 V and that across R is 100 V. Calculate the p.d. across the inductor.

SOLUTION 48

From the voltage triangle of Fig. 148

$$V^2 = V_L^2 + V_R^2$$
$$240^2 = V_L^2 + 100^2$$
$$V_L^2 = 240^2 - 100^2$$
$$= 57,600 - 10,000$$
$$= 47,600$$
$$V_L = 218.2 \text{ V.}$$

WORKED EXAMPLE 49

In the circuit shown in Fig. 154 Calculate (i) the supply current (ii) the supply phase angle (iii) the p.ds across C and R.

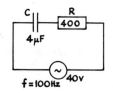

Fig. 154   R-C series circuit

SOLUTION 49

(i) $Z = \sqrt{R^2 + X_C^2} = \sqrt{400^2 + 398^2} = 564.2 \ \Omega$

where $X_C = \dfrac{1}{2 \pi f C} = \dfrac{1}{2 \pi \times 100 \times 4 \times 10^{-6}} = 398 \ \Omega$

$I = \dfrac{V}{Z} = \dfrac{40}{564.2} = 0.071 \text{ A.}$

(ii) $\tan \phi = \dfrac{X_C}{R} = \dfrac{398}{400} = 0.995$ from the

impedance triangle of Fig. 153, $\phi = 44.9°$ is the phase angle

(iii) $V_C = I X_C = 0.071 \times 398 = 28.3 \text{ V}$

$V_R = I R = 0.071 \times 400 = 28.4 \text{ V.}$

g.   *Shows graphically that where currents and voltages are sinusoidal:*
   * *for a purely resistive a.c. circuit, average power is VI*
   * *for a purely reactive a.c. circuit, average power is zero*
   * *for a resistive/reactive a.c. circuit, average power depends upon phase-angle.*

POWER IN A PURELY RESISTIVE A.C. CIRCUIT

At any instant, t, the power is iv, $p = iv$, i and v are in phase. Fig. 136 shows the circuit and the relevant waveforms.

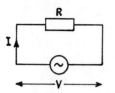
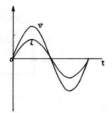

Fig. 136   Circuit and i, v waveforms.

The current through R and the voltage across R are in phase as shown, the r.m.s. current is I and the r.m.s voltage is V, the average power $P = IV = V^2/R = I^2R$.

POWER IN A PURELY REACTIVE A.C. CIRCUIT

Fig. 155 and Fig. 156 show the circuits and relevant i, v waveforms.

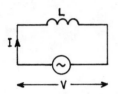
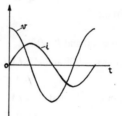

Fig. 155   The pure inductor and i, v_L, waveforms.

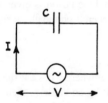
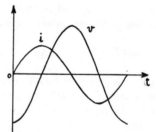

Fig. 156   The pure capacitor and i, v_L, waveforms.

At any instant, t, the power is iv, $p = iv$, i and v are out of phase by 90° when $i = 0$, v_L is maximum when $v_L = 0$, i is maximum. Similarly for the capacitor when $i = 0$, v_C is negative maximum, when i is maximum v_C is zero and the average power is zero.

POWER IN A RESISTIVE/REACTIVE A.C. CIRCUIT

The average power in a resistive/reactive circuit depends on the phase angle.

$P = IV \cos \phi$.

Resistive/reactive a.c. circuits are shown in Fig. 157 and Fig. 158.

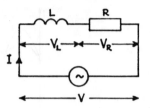

Fig. 157   *LR* circuit

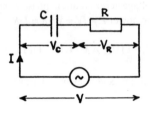

Fig. 158   *CR* circuit

For a pure resistor $\phi = 0$ and $P = IV$.

For a pure capacitor $\phi = 90°$ and $P = I\,V \cos 90°$
$$= 0.$$

For a pure inductor $\phi = 90°$ and $P = I\,V \cos 90°$
$$= 0.$$

In an *LR* and *RC* circuits *I* and *V* normally have a phase angle ϕ and the Power is given by *I V* cos ϕ. Therefore the average power in a resistive/reactive circuit depends on the phase angle.

h.   *States that $P = V\,I \cos \phi$ for sinusoidal waveforms.*

If *V* leads *I* by an angle ϕ then $P = I\,(V \cos \phi) = V\,I \cos \phi$ active power where *V* cos ϕ is the active component of voltage. Fig. 159 shows this situation. If *I* leads *V* by an angle ϕ, then $P = V\,(I \cos \phi) = V\,I \cos \phi$ active power where *I* cos ϕ is the active component of current Fig. 160 shows this situation.

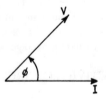

Fig. 159   *V* leads *I* by an angle ϕ.

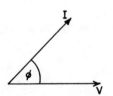

Fig. 160   *I* leads *V* by an angle ϕ.

i. Derives the power triangle from the voltage triangle and identifies true power (P), apparent power (S), and reactive volt amperes (Q).

From the voltage triangles of Fig. 148 and Fig. 151, we multiply each side by I, we have the power triangles as shown in Fig. 161 and Fig. 162 respectively.

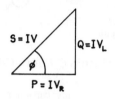

Fig. 161   Power triangle for inductive circuit

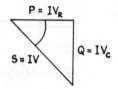

Fig. 162   Power triangle for capacitive circuit.

Q = reactive volt amperes = $I\,V \sin \phi$

S = apparent power = IV

P = true power = $IV \cos \phi$.

j. Defines power factor as true power/ apparent power and show that where V and I are sinusoidal, power factor = cos ϕ.

POWER FACTOR

Power factor $= \dfrac{\text{true power}}{\text{apparent power}}$

$\qquad = \dfrac{I\,V \cos \phi}{IV}$

$\qquad = \cos \phi$.

k. Explains that power dissipation in series L-R amd C-R a.c. circuits is I^2R.

POWER DISSIPATION IN SERIES *L-R* A.C. CIRCUITS

The power dissipation in a pure inductor is zero and hence the only power which is dissipated is in the resistor R. This power is $P = I^2R$. Fig. 163 shows in L-R a.c. circuit.

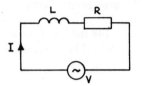

Fig. 163   Power in *L-R* a.c. circuit.

POWER DISSIPATION IN SERIES *C-R* A.C. CIRCUIT

Similarly the power dissipation in a pure capacitor is zero and the only power which is dissipated is in the resistor R. Again this power is $P = I^2R$. Fig. 164 shows the R-C a.c. circuit.

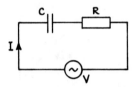

Fig. 164   Power in *R-C* a.c. circuit.

l.   Uses phasor diagrams to solve simple series L,C and R a.c. circuits.

R-L-C SERIES AC CIRCUIT

Fig. 165 shows the series *RLC* a.c. circuit subject to an a.c. supply of r.m.s. voltage V and r.m.s. current I.

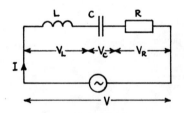

Fig. 165   *R-L-C* series circuit.

There are three cases to consider

a.   If $V_L > V_C$

b.   If $V_L < V_C$

c.   If $V_L = V_C$.

The corresponding phasor diagrams are shown in Figs. 166, 167 and 168.

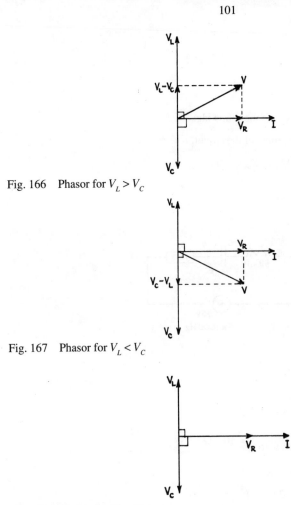

Fig. 166   Phasor for $V_L > V_C$

Fig. 167   Phasor for $V_L < V_C$

Fig. 168   Phasor for $V_L = V_C$

From these phasor diagrams, we have the impedance triangles shown in Fig. 169 and Fig. 170 corresponding to Fig. 166 Fig. 167, Fig. 168 has no impedance triangle, $Z = R$.

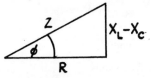

Fig. 169   Impedance triangle for the inductive case.

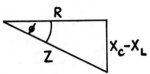

Fig. 170   Impedance triangle for the capacitive case.

From Fig. 169 we have

$$Z = \sqrt{R^2 + (X_L - X_C)^2}$$

From Fig. 170 we have

$$Z = \sqrt{R^2 + (X_C - X_L)^2}.$$

WORKED EXAMPLE 50

In the circuit shown in Fig. 171 calculate the following:

(i)   The total impedance and phase angle of the circuit.

(ii) The voltage across the capacitor.

(iii) The voltage across the coil.

(v)   The power developed in the circuit.

(vi) The power factor of the circuit.

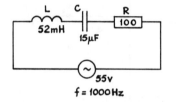

Fig. 171   A series *RLC* a.c. circuit.

SOLUTION 50

(i) $X_L = 2\pi fL = 2\pi \times 1{,}000 \times 52 \times 10^{-3} = 326.7\ \Omega$

$X_C = \dfrac{1}{2\pi fC} = \dfrac{1}{2\pi \times 1000 \times 15 \times 10^{-6}} = 10.6\ \Omega$

From the impedance triangle of Fig. 169

$Z = \sqrt{100^2 + (326.7 - 10.6)^2} = 331.5\ \Omega$

using the equation $Z = \sqrt{R^2 + (X_L - X_C)^2}$.

(ii) $V_C = IX_C = 0.166 \times 10.6 = 1.76$ volts

where $I = \dfrac{V}{Z} = \dfrac{55}{331.5} = 0.166$ A.

(iii) $V_R = IR = 0.166 \times 100 = 16.6$ volts

(iv) $V_L = IX_L = 0.166 \times 326.7 = 54.23$ volts

(v) $P = I^2R = 0.166^2 \times 100 = 2.76$ watts

(vi) $\cos\phi = \dfrac{R}{Z} = \dfrac{100}{331.5} = 0.302.$

WORKED EXAMPLE 51

For the circuit shown in Fig. 172.

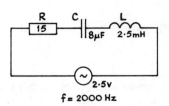

Fig. 172   Series *RLC* a.c. circuit.

Calculate: (i) the impedance,

(ii) the total current,

(iii) the voltage across each component,

(iv) the power factor,

(v) the power dissipated.

SOLUTION 51

(i) $X_L = 2\pi fL = 2\pi \times 2,000 \times 2.5 \times 10^{-3} = 31.42\ \Omega$

$X_C = \dfrac{1}{2\pi fC} = \dfrac{1}{2\pi \times 2,000 \times 8 \times 10^{-6}} = 9.95\ \Omega$

$Z = \sqrt{R^2 + (X_L - X_C)^2} = \sqrt{15^2 + (31.42 - 9.95)^2} = 26.2\ \Omega$

(ii) $I = \dfrac{V}{Z} = \dfrac{2.5}{26.2} = 0.4\ A$

(iii) $V_R = 0.4 \times 15 = 0.6\ V$

$V_L = 0.4 \times 31.42 = 12.6\ V$

$V_C = 0.4 \times 9.95 = 3.98\ V$

(iv) $\cos\phi = \dfrac{R}{Z} = \dfrac{15}{26.2} = 0.573$

(v) $P = I^2R = 0.4^2 \times 15 = 2.4\ W.$

m.   Defines series resonance as accuring when the supply voltage and current are in phase and sketches a phasor diagram showing that:

- $V_L = V_c$ *at resonance*
- V_L *and* V_c *may be much greater than the supply voltage.*

SERIES RESONANCE

Series resonance occurs when the supply voltage and current are in phase.

Fig. 173 shows this condition.

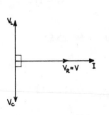

Fig. 173   Series resonance.

At resonance

$V_L = V_C$ and the supply voltage $V = V_R$

$$I_o \, 2\pi f_o L = I_o \frac{1}{2\pi f_o C}$$

where I_o is the current in the circuit at resonance which is maximum and the frequency f_o is the resonant frequency from equation (1) we have

$$f_o = \frac{1}{2\pi \sqrt{LC}}$$

$$V_L = I_o X_L = \frac{V}{R} X_L = V \left(\frac{X_L}{R}\right)$$

$$V_c = I_o X_c = \frac{V}{R} X_c = V \left(\frac{X_c}{R}\right).$$

The ratio $\dfrac{X_L}{R}$ or $\dfrac{X_C}{R}$ may be much

greater than unit, and hence the p.d. across X_L or X_C is much greater than the supply voltage, V.

WORKED EXAMPLE 52

A series circuit consists of a non-inductive resistance of 20 Ω, a capacitance of 10 μF, and pure inductance of 150 mH. The a.c. supply voltage is 200 V at 100 Hz.

Calculate: (i) the circuit impedance

           (ii) the current in the circuit

           (iii) the phase angle.

Connect another component in series with the above in order that the impedance of the circuit is only 20 Ω or the phase angle is zero or the current in the circuit is maximum. Determine the new value of current.

SOLUTION 52

Fig. 174 shows the series LRC a.c. circuit.

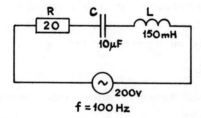

Fig. 174 LRC series circuit.

(i) $X_L = 2\pi fL = 2\pi \times 100 \times 150 \times 10^{-3} = 94.25 \ \Omega$

     $X = 94.25 \ \Omega$

     $X_c = \dfrac{1}{2\pi fC} = \dfrac{1}{2\pi \times 100 \times 10 \times 10^{-6}} = 159.2 \ \Omega$

     $X_C = 159.2 \ \Omega$

$$Z = \sqrt{20^2 + (159.2 - 94.25)^2} = 67.96\ \Omega$$

(ii) $I = \dfrac{V}{Z} = \dfrac{200}{67.96} = 2.94\ \text{A}$

(iii) $\tan \phi = \dfrac{X_C - X_L}{R} = \dfrac{159.2 - 94.25}{20} = 3.25$

  $\phi = 72.9°.$

The other component to be added must be a pure coil of reactance $159.2 - 94.25 = 65$ approximately

$$X_L = 2\pi fL = 65$$

$$L = \dfrac{65}{2\pi \times 100} = 103\ \text{mH}$$

such that $X_L = X_C$

$$I_o = \dfrac{V}{R} = \dfrac{200}{20} = 10\ \text{A}.$$

This problem is only a theoretical exercise since there is no pure coil in practice. Fig. 175 shows the new circuit for resonance.

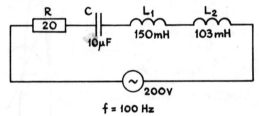

f = 100 Hz

Fig. 175   103 mH is added to cause resonance.

WORKED EXAMPLE 53

A series circuit consists of a coil of inductance of 250 mH, a resistor of resistance 30 Ω, and a capacitor of capacitance C. The circuit resonates when the supply is 240 V and the power frequency is 50 Hz.

Calculate:

  (i) the value of C

  (ii) the resonant current

  (iii) the voltage across the inductor

  (iv) the power dissipated

  (v) the Q-factor of the circuit

  where $Q = 2\pi f_o L / R$.

SOLUTION 53

(i) $X_L = 2\pi f_o L = 2\pi \times 50 \times 250 \times 10^{-3} = 78.54\ \Omega$

  $X_L = 78.54\ \Omega$

$$X_C = 78.54 = \frac{1}{2\pi f_o C}$$

$$C = \frac{1}{2\pi \times 50 \times 78.54} = 40.5\ \mu F$$

(ii) $I_o = V/R = 240/30 = 8A$

(iii) $V_L = I_o X_L = 8 \times 78.54 = 628.32$ volts

(iv) $P = I_o^2 R = 8^2 \times 30 = 64 \times 30 = 1{,}920$ W

(v) $Q =$ quality factor $= \omega_o L/R = \dfrac{78.54}{R} = 2.618$

$$Q = \frac{628.32}{240} = \frac{V_L}{V} = 2.618$$

which verifies that V_L is equal to Q times the supply voltage V.

SUMMARY 7

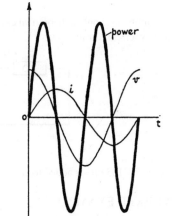

PURE RESISTOR $R = \dfrac{V}{I}$ (Ω) ohm's law

PURE INDUCTOR $X_L = \dfrac{V}{I}$ (Ω) ohm's law

PURE CAPACITOR $X_C = \dfrac{V}{I}$ (Ω) ohm's law

INDUCTIVE REACTANCE $X_L = 2\pi fL$ (Ω)

CAPACITIVE REACTANCE $X_C = \dfrac{1}{2\pi FC}$ (Ω)

SERIES INDUCTIVE RL CIRCUIT $V^2 = V_R^2 + V_L^2$

SERIES CAPACITIVE RC CIRCUIT $V^2 = V_R^2 + V_C^2$

IMPEDANCES $Z = \sqrt{R^2 + X_L^2}$ $\qquad Z = \sqrt{R^2 + X_C^2}$

IMPEDANCE TRIANGLE $\tan\phi = \dfrac{X}{R}$ $\quad \sin\phi = \dfrac{X}{Z}$

$$\cos\phi = \frac{R}{Z}$$

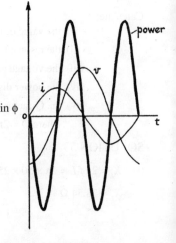

POWER TRIANGLE $Q =$ REACTIVE VOLTAMPERES $= IV \sin\phi$

$$S = \text{APPARENT POWER} = IV$$

$$P = \text{TRUE POWER} = IV \cos\phi$$

RESISTIVE POWER $= P = I^2 R = IV = \dfrac{V^2}{R}$

RLC CIRCUIT $Z = \sqrt{R^2 + (X_L - X_C)^2}$

RESONANCE: $Z = R$ $f_o = \dfrac{1}{2\pi \sqrt{LC}}$ $I_o = \dfrac{V}{R}$

$$V_L = QV$$

$$V_C = QV$$

$$Q = \dfrac{X_L}{R} \qquad\qquad Q = \dfrac{X_C}{R}.$$

EXERCISE 7

1. A capacitor of 100 μF capacitance is subjected to an a.c. supply of frequency 100 Hz. Determine the capacitive reactance. (Ans. 15.9 Ω).

2. An inductor of 50 mH inductance is subjected to an a.c. supply of frequency 60 Hz. Determine the inductive reactance. (Ans. 18.9 Ω).

3. The reactance of a capacitor is 100 Ω when subjected to an a.c. supply of 50 V. Calculate the current through the circuit. (Ans. 0.5 A).

4. The inductive reactance of a coil is 5 Ω when it is subjected to an a.c. supply which draws 5 A current. Calculate the a.c. supply. (Ans. 25 V).

5. Sketch neatly a sine wave of current, and in relation to this current sketch the voltage waveform you would expect when this current passes through (i) a resistance (ii) an inductance (iii) a capacitance. Label your diagrams.

6. A series circuit consists of an inductor of inductance 0.5 H and a resistor of resistance 50 Ω. A 240 V a.c. supply of frequency of 50 Hz is connected across the above circuit. Calculate (i) the impedance of the circuit (ii) the current drawn from the a.c. supply (iii) the p.d. across the pure inductor (iv) the p.d. across the resistor and hence check that the supply voltage is 240 volts.

   (Ans. (i) 164.9 Ω, (ii) 1.46 A (iii) 229.3 V (iv) 73 V).

7. A coil takes a current of 5 A when connected to a 100 V d.c. supply, when the same coil is connected to a 100 V 50 Hz a.c. supply the current drawn is 1 A. Explain why these currents differ.

   Calculate: (i) the resistance of the coil

                   (ii) the impedance of the coil at 50 Hz.

   (Ans. 20 Ω, 100 Ω).

8. An inductor takes 10 A and dissipates 500 W when connected to a 200 V 50 Hz a.c. supply.

   Calculate: (i) The impedance,

                   (ii) the power factor,

                   (iii) the resistance,

                   (iv) the inductance.

   (Ans. (i) 20 Ω (ii) 0.25 (iii) 5 Ω (iv) 61.6 mH).

9. The power taken by an inductive circuit when connected to a 240 V, 50 Hz supply is 1000 watts, and the current is 16 A. Calculate;   (i) The power factor of the circuit,

(ii) the impedance,

(iii) the resistance,

(iv) the reactance,

(v) the inductance.

(Ans. (i) 0.26 (ii) 15 Ω (iii) 3.9 Ω (iv) 14.48 Ω (v) 46.1 mH.)

10. An RLC series circuit has a resistance of 15 Ω, an inductance of 300 mH and a capacitance of 100 μF. Calcualte:

(i) The resonant frequency.

(ii) The supply current.

(iii) The p.d. across the inductance.

(iv) The p.d across the capacitance.

(v) The power at resonance.

Given that the applied a.c. voltage is 240 V.

(Ans. (i) 29.06 Hz (ii) 16 A (iii) 876.4 V (iv) 876.4 V (v) 3,840 W.)

11. A variable inductor with negligible effective resistance is connected in series with a resistor and a capacitor. The whole circuit is applied across an a.c. supply of 240 V, 50 Hz, the inductor was adjusted to give a maximum current of 1 A.

The p.d. across the capacitor is 500 volts at this maximum current. Calculate the value of the resistance, the capacitance and inductance.

(Ans. 240 Ω, 6.37 μF and 1.59 H).

12. A series circuit consists of a coil of resistance 50 Ω and inductance 0.05 H, and a capacitor 0.2 μF. The circuit is subjected to an a.c. supply of 300 V. Calculate:

(i) The resonant frequency.

(ii) The resonant current.

(iii) The p.d across the capacitor.

(iv) The quality factor, Q.

(Ans. (i) 1591.6 Hz (ii) 6 A (iii) 3,000 V (iv) 10).

13. An inductive circuit of resistance 10 Ω and inductance 50 mH is connected in series with a capacitor. When connected across a 240 V 50 Hz supply, the current taken is 5 A leading. What is the value of the capacitor? Repeat the calculation if the current taken is 2 A lagging.

(Ans. 15.6 μF, 84.8 μF).

MEASURING INSTRUMENTS AND MEASUREMENT

8. *Understands the operation and limitations of measuring instruments and uses them correctly in a wide range of basic applications.*

    a. *Describes with the aid of given diagrams, the principles of operation of moving coil instruments and explains the need for shunts and multipliers to extend the range of basic electrical indicating instruments.*

MOVING COIL INSTRUMENT

This instrument is basically a coil of fine insulated copper wire of N turns wound on a rectangular former as shown in Fig. 177.

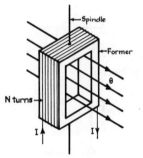

Fig. 177   Coil of N turns wound on a rectangular former.

This rotates about a central spindle in strong magnetic field. The currents flowing through the narrower sides are parallel with the field B and the forces exerted are zero. The current flowing through the longer sides are at right angles and therefore the forces are maximum and equal to $F = BIl$ where l = length of longer sides.

A soft iron cylinder is placed as shown in Fig. 178 and together with a permanent magnet produces a radial field as shown in Fig. 179.

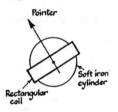

Fig. 178   Rectangular coil on soft iron cylinder.

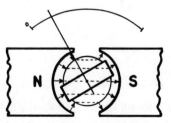

Fig. 179   Rectangular coil place in a permanent magnet.

A pointer is attached at right angles to the spindle.

The coil as pivoted on jewelled bearings and its rotation is resisted by hair springs above and below the coil.

Each of the vertical conductors (long sides) of the coil is placed right angles to the radial field produced by the permanent magnets.

If the Left Hand Rule is applied, the following situations occur which are shown in details in figs 180, 181 and 182.

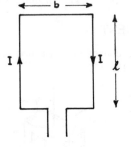

Fig. 180   One turn of the rectangular coil.

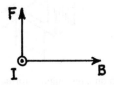

Fig. 181   Force coming out of the paper for the left hand side of the rectangular coil.

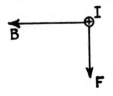

Fig. 182   Force going into the paper for the right hand side of the rectangular coil.

A couple of forces occur as in Fig. 183

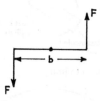

Fig. 183   Couple of forces $F \times b$

and the coil rotates about the central spindle.

CURRENT SENSITIVITY

Couple of forces is defined as $F \times b = BIl \times N \times b = c\,\theta$ where c is the controlling constant and θ is the deflection of the pointer in radians

$$\theta c = BIlNb$$

$$\frac{\theta}{I} = \frac{BNA}{c} \quad \text{where } A = bl \text{ the area of the rectangle}$$

$$\frac{\theta}{I} = \text{deflection per unit current}$$
$$= \text{current sensitivity}$$

For maximum current sensitivity, B should be large, N is large, and A is large and c must be as small as it is possible.

THE SHUNT

A galvanometer is a sensitive ammeter, measuring very small currents. In order to obtain larger currents, a resistor called the *shunt* is connected in parellel with the galvanometer.

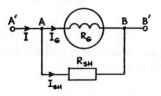

Fig. 184   The shunt.

$$I_{sh} = \frac{R_G}{R_G + R_{sh}} I$$

$$I_G = \frac{R_{sh}}{R_G + R_{sh}} I$$

$$I = I_G + I_{sh} = I_G + \frac{R_G}{R_{sh}} = I_G + I_G \left(\frac{R_{sh} + R_G}{R_{sh}} \right)$$

$$\frac{I}{I_G} = \frac{R_{sh}}{R_{sh}} + \frac{R_G}{R_{sh}} = 1 + \frac{R_G}{R_{sh}} \text{ or } \frac{R_{sh}}{R_G} = \frac{1}{\dfrac{I}{I_G} - I}$$

$$R_{sh} = \frac{R_G}{\left(\dfrac{I - I_G}{I_G} \right)} .$$

The terminals of the extended scale instrument, are, A' and B'.

THE MULTIPLIER

A *multiplier* is a large resistance R_M connected in series with the galvanometer in order to extend the voltage scale. Fig. 185 shows the principle of multiplier.

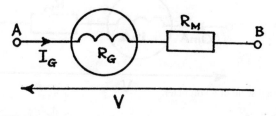

Fig. 185   The multiplier.

From Fig. 185 we have

$$V = I_G (R_G + R_M)$$

re-arranging the formula for R_M.

$$R_M = \frac{V}{I_G} - R_G.$$

The new voltmeter terminals are now A and B.

WORKED EXAMPLE 54

A moving coil instrument has a resistance of 15 Ω and the full scale deflection is 50 mA. How would you adapt this instrument for use as an ammeter reading 0-5 amperes?

SOLUTION 54

In order to adapt the moving coil instrument to read full scale deflection of 5 A, a shunt is connected across the instrument as shown in Fig. 186.

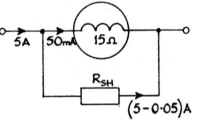

Fig. 186   Extended scale of am ammeter.

$$I_{sh} = \frac{15}{15 + R_{sh}} \ 5 = 4.95$$

$$15 + R_{sh} = \frac{15 \times 5}{4.95} = 15.151515$$

$$R_{sh} = 0.151515.$$

WORKED EXAMPLE 55

A moving coil instrument has a resistance of 15 Ω and the full scale deflection is 50 mA. How would you adapt this instrument for use as a voltmeter reading 0–10 V?

SOLUTION 55

In order to adapt the moving coil instrument to read full scale deflection of 10 V, a multiplier resistor is connected in series as shown in Fig. 187.

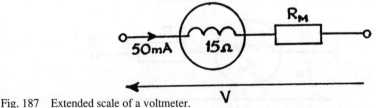

Fig. 187   Extended scale of a voltmeter.

From Fig. 187 we have

$$10 = 50 \times 10^{-3} \times 15 + R_M \times 50 \times 10^{-3}$$

$$10{,}000 = 50\,(15 + R_M)$$

$$15 + R_M = 200$$

$$R_M = 185\ \Omega.$$

b. *Uses ammeters and voltmeters correctly in d.c. circuit measurement.*

A BASIC AMMETER

A moving coil instrument is basically a d.c. ammeter and is used to detect current from a few mA (milliamperes) to several hundred amperes. Galvanometers are used to measure small currents, only in the range of μA (micro-amperes). They have a large number of turns, are very sensitive and delicate instruments. Galvanometers have large resistances in the order of 50 Ω or 100 Ω but in conjunction with shunts, the resistance is greatly reduced. The higher the full scale deflection (f.s.d.) the smaller is the resistance between the terminals. Therefore, ammeters normally have a *very small resistance* between their terminals.

Ammeters are connected in series with the component whose current is required to be measured.

A BASIC VOLTMETER

A d.c. voltmeter is normally a moving coil d.c. ammeter with a large resistor connected in series and thus voltmeters normally have a very large resistance between their terminals.

Voltmeters are connected in parallel with the component whose voltage is required to be measured.

AMMETER AND VOLTMETER METHOD FOR MEASURING RESISTANCE

There are two cases to consider:

(a) the measurement of a low value of resistance of the order of 10 Ω and (b) the measurement of a high value of resistance of the order of 5,000 Ω.

Fig. 188 shows the connection of the ammeter for low values of resistance.

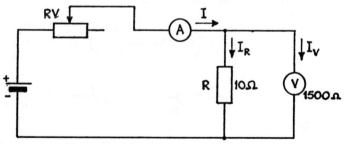

Fig. 188 Measurement of low values of resistance.

The current indicated on the ammeter is the total current of the circuit.

If the resistance of the voltmeter, $R_V = 1{,}500\ \Omega$, then the currents I_R and I_V are given by the current divider as

$$I_R = \frac{1{,}500}{1{,}510}\,I = 0.9934\,I \quad \dots (1)$$

$$I_V = \frac{10}{1,510} \; I = 0.0066 \, I \quad \dots (2)$$

In order to measure the resistance, R, a set of readings of I and V are noted as the rheostat is varied, R is then calculated from the average value of the ratios of V/I, which will be slightly less than the actual value.

The percentage error in this case will be about 0.66. This is shown in equation (2) and I_R is taken to be approximately equal to the total current I.

Fig. 189 shows the connection of the ammeter for high values of resistance.

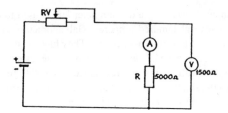

Fig. 189   Measurement of high values of resistance.

The current indicated on the ammeter is the actual current that flows through the resistance which is to be measured. But the voltmeter reading indicated will be slightly higher, since it will be measuring the p.d. of the ammeter as well, so in this case the value of resistance R will be slightly higher by considering the ratio of V/I.

In both the above methods, however, there are small errors.

c.   *Describes with the aid of diagrams the principles of operation of an ohm meter and uses an ohm meter for the measurement of resistance.*

THE BASIC PRINCIPLE OF AN OHM METER

A moving coil is connected in series with a variable resistor R, a battery of 3 V and the two terminals A and B as shown in Fig. 190. When A and B are short circuited and R is adjusted for full scale deflection of current, then this corresponds to zero resistance between A and B.

The scale of ohmmeter is calibrated as shown in Fig. 191.

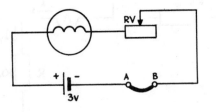

Fig. 190   The ohmmeter.

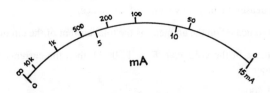

Fig. 191   Calibrated scale of an ohmmeter is non linear.

The short circuit is now replaced with standard resistors of different values such as 100 Ω, 500 Ω, 1,000 Ω, 10,000 Ω and they are placed in turn and then the resistance scale is calibrated as shown in Fig. 191. Note that the scale will be non-linear.

To measure an unknown resistance R_x, it is placed between A and B terminals thus giving a deflection which gives its value.

USE OF AN OHMMETER

Short circuit A and B and adjust R to give zero resistance. Zero (corresponds to f.s.d. of current). Then replace the short circuit with the unknown resistance and read off its value from the calibrated scale.

d.   Explains the need for rectifier instruments and states the frequency and waveform limitation inherent in moving-coil rectifier instruments.

RECTIFIER INSTRUMENT

A moving-coil instrument is used in conjunction with a rectifier bridge circuit in order to provide an indication of alternating quantites. This arrangement is shown in Fig. 192.

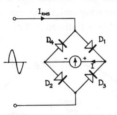

Fig. 192   Rectifier instrument.

When A is positive with respect to B, current flows through D_1, through the moving coil meter, and D_2.

When B is positive with respect to A, current flows through D_3, the moving coil meter and D_4.

The d.c. current, $I_{d.c.}$ flowing through the M.C., and the a.c. current I are connected by the expression

$$I_{d.c.} = \frac{I_{d.c}}{1.11}$$

where 1.11 as the form factor of the sinusoidal waveform applied as the a.c. signal.

Rectifier instruments have their scales calibrated in terms of r.m.s. quantities and the a.c. signals are sinusoidal.

These instruments may be frequency limited. Rectifier moving coil instruments can be used on frequencies up to several KHz.

e.   Uses electronic and moving - coil multimeter correctly for the measurements of I and V in d.c. and a.c. circuits.

We have seen previously that voltmeters possess much higher resistances between their terminals, than ammeters.

Voltmeters are therefore graded according to their 'resistance per volt' at full scale deflection (f.s.d.). The total resistance of the meter is that of the coil and the resistance of the multiplier. If the full scale deflection is 100 μA amd the p.d. is 1 V between the terminals then the total resistance, R, is given by

$$R = \frac{1}{100 \times 10^{-6}} = \frac{10^6}{100} = 10,000 \ \Omega$$

and the *resistance per volt* is 10,000 Ω/V. A good grade voltmeter must have a resistance of at least 1,000 Ω/V.

Electronic voltmeters are widely used where the resistance between the terminals is of the order of 10 MΩ and thus the current through the meter is negligible. Thus for accurate measurements electronic voltmeters are preferred for d.c. and a.c. circuits.

f.   Uses a wattmeter

THE DYNAMOMETER WATTMETER

The dynamometer wattmeter is the most commonly used instrument in a.c. circuits. The dynamometer instrument is a moving coil instrument in which the magnetic field is not provided by a permanent magnet, but by another pair of coils. The dynamometer principle is basically made up of two fixed field coils and a moving coil, the current passes through the fixed coils and sets up a magnetic field which is effectively uniform in the region around the moving coil. This is illustrated in the Fig. 193.

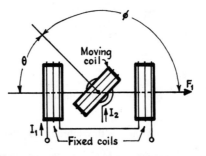

Fig. 193

The magnetic field is produced by the current I_1 flowing in two fixed circular coils which are connected in series in this case (then can be connected in parallel also). The field between the two coils is substantially uniform, and the moving coil, also circular, and carrying current I_2 rotates in this field.

Therefore when a dynamometer instrument is used as a wattmeter the currents in the fixed and moving coils are different.

The fixed coil curry the load current, and the moving coil carries a current proportional to the voltage across the load.

The m.m.f.'s F_1 and F_2 act in the directions indicated, and the operating torque T is approximately given by

$$T = KF_1 F_2 \sin \theta$$

since F_1 is proportional to I_1 and F_2 to I_2, and sin θ in numerically equal to sin θ

$T = KI_1 I_2 \sin \theta$

when the instrument is used as an ammeter the moving coil is connected either in series or in parallel with the fixed coils, I_1 and I_2 are then both proportional to the current I being measured, so the torque is proportional to I^2, and the instrument reads true r.m.s.

The dynamometer instrument can thus be used on alternating current circuits. The wattmeter connection on a single-phase circuit are usually made as shown in Fig. 194.

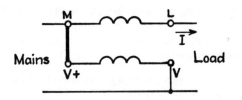

Fig. 194   wattmeter connection for measurement of power on a single-phase system. Normal connection.

Using these connections, the current coils measure the true current taken by the load, but the voltage coils measure the p.d. across the current coil in addition to the p.d. across the load. The result is that the power loss in the current coils of the instrument is included in the indication. The error is usually small, but a compensation can be made if the resistance r_c of the current coils is known. The power loss $I^2 r_c$ is then calculated and subtracted from the wattmeter reading.

Alternative connections are shown in Fig. 195.

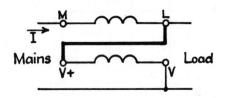

Fig. 195   wattmeter connections for measurement of power on a single-phase system Alternative connections.

The voltage coil then measures the voltage across the load, but the current coil measures the current taken by the voltage coil in addition to the current taken by the load. The result is that the power loss in the voltage coil is indicated in the reading. Again the error is usually small, but compensation can be made if the resistance r_v of the voltage coil is known. The power loss V^2/r_v can then be calculated and subtracted from the wattmeter reading.

g.   *Uses double beam CRO in direct and indirect measurement of a.c. and d.c. voltages and to measure period and frequency.*

CATHODE RAY OSCILLOSCOPE (C.R.O.)

A cathode Ray Oscilloscope (C.R.O) with two amplifiers (double beam) is used to measure voltages visually. That is, two waveforms may be displayed on the C.R.O. at any one time and thus the alternating voltage waveforms are studied. A special electronic circuit is built within the C.R.O and is called *the time base*. The time base circuit produces an output voltage with the saw-tooth waveform as shown in Fig. 196.

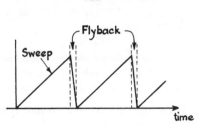

Fig. 196   Time base waveform.

The time base period = the sweeping time + the flyback time. This voltage is connected across the × plates in such a way that the spot is scanned across the screen from left to the right relatively slowly and is then flies back to the left hand side very quickly. This cycle then is repeated. The voltage to be examined is connected across the Y-plates.

If the frequency of the time base is adjusted to be the same as the frequency on the y-plates, a steady waveform will be displayed on the screen.

THE CATHODE RAY TUBE (C.R.T.)

The cathode ray tube is the heart of the C.R.O. and it is simply described here.

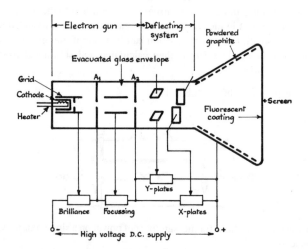

Fig. 197   Construction of an Electrostatic Cathode Ray Tube. (C.R.T.).

The C.R.T. is used to study the waveforms of d.c. and a.c. electrical voltages and currents. The construction of a typical C.R.T. with electrostatic deflection and focusing is shown in Fig. 197. It consists basically of the electron gun, the deflecting system of plates and the fluorescent screen.

An indirectly heated cathode in the form of a nickel cylinder with an oxide coated end cap. A cylindrical grid, houses the cathode with negative potential for controlling the emitted electrons from the cathode. The anodes A_1 and A_2 are discs with central, holes through which the electrons pass, a cylindrical anode, the focusing anode, is placed between A_1 and A_2. The beam passes between two pairs of parallel deflecting plates, which are mutually perpendicular, the y-plates and the x-plates. A p.d. across the x-plates produces a horizontal deflection and a p.d. across the y-plates produces vertical deflection of the spot on the screen.

The funnel-shaped part of the c.r.t. is coated internally with graphite and is earthed. This allows the electrons which strike the screen to leak away. The screen however is coated with a fluorescent powder, such as zinc sulphide (blue glow) or $Z_{in}C$ orthosilicate (blue-green glow).

The C.R.O. has a high input impedance which means that it does not draw a high current. The C.R.O. is a visual electronic voltmeter measuring a.c. and d.c. waveforms.

To measure a voltage, the peak-to-peak distance of the trace is measured in centimetres using the graticule mounted in front of the screen. Knowing the sensitivity in volts per centimetre then enables us to convert this reading into volts. The amplifier controls are calibrated in V/cm or mV/cm, these calibrations are approximate.

PERIOD AND FREQUENCY MEASUREMENT X AND Y CONTROLS

The time base switch can be adjusted to measure the period and hence the frequency of the waveform. Let us assume that one complete cycle of a sinusoidal waveform is displayed on the screen, and that the length of the horizontal distance is 3.6 cm an the time base control is set at 100 μs/cm. The periodic time of the waveform is given by 3.6 cm × 100 μs/cm = 360 μs (provided the C.R.O. is calibrated).

$$T = 360 \times 10^{-6}\,\text{s}.$$

The frequency of the waveform is found to be

$$f = \frac{1}{1} = \frac{1}{360 \times 10^{-6}}\,\text{Hz}$$

$$= 2,778\,\text{Hz}$$

$$f = 2.78\,\text{KHz}$$

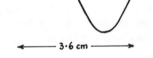

Fig. 198   Period and frequency measurement of a sinusoidal waveform.

WORKED EXAMPLE 56

A sinusoidal waveform is displayed on the screen of an oscilloscope and the following measurements where made on the graticule of the screen:

Peak - to - peak distance = 5.7 cm (vertical) horizontal distance between the two peaks = 4.8 cm. The Y amplifier setting is 500 mV/cm and the time base setting is 100 μs/cm.

Determine the approximate r.m.s. voltage, period and frequency.

SOLUTION 56

Peak-to-peak voltage= 5.7 cm × 500 × 10^{-3} V/cm

$$= 2.85\,\text{V}$$

Peak voltage = 1.43 V

$$\text{r.m.s. voltage} = \frac{1.43}{\sqrt{2}} = 1 \text{ V approximately}$$

$$\text{Periodic time} = 4.8 \text{ cm} \times 100 \times 10^{-6} \text{ s/cm}$$

$$= 0.48 \text{ ms}$$

$$\text{Frequency} = \frac{1}{T} = \frac{1}{0.48 \times 10^{-3}} = 2{,}083 \text{ Hz approximately}$$

LISSAJOU'S FIGURES

A more accurate way of measuring frequency is to use the C.R.O to make a comparison between the unknown frequency and a signal of known frequency. The known frequency is derived from a calibrated signal generator (oscillator). The unknown signal is applied to the Y plates and the output from the signal generator to the X plates, the time base being switched off. When the frequency of the signal generator is suitably adjusted, Lissajon's figures can be formed on the C.R.O. screen. The shape of the figure depends on the ratio of the two frequencies. For simple ratios such as 1:1 and 2:1 the figures are an ellipse and a figure of eight respectively.

h.   Describes the principle of null method of measurement.

A sensitive ammeter or galvanometer is used with a centre zero (cz) as shown in Fig. 199.

Fig. 199   A centre zero galvanometer instrument.

The pointer is at the centre of the instrument and may deflect to the left or to the right from the centre, thus indicating that the current either flows from left to right or from right to left. A null method is a method which indicates that the current through the instrument is zero so that the p.d. across the terminals is zero, and thus the power loss in the instrument is negligible.

i   Describes with the aid of diagram the principle of the Wheatstone bridge and uses a Wheatstone bridge to measure resistance over a wide range.

WHEATSTONE BRIDGE

Four resistors are placed as shown in Fig. 200 forming a bridge.

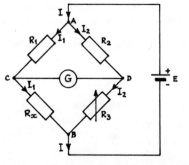

Fig. 200   Wheatstone bridge.

The d.c. supply is connected across one diagonal and the sensitive ammeter or galvanometer across the other diagonal. R_1 and R_2 are two standard resistors of high accuracy, R_3 is a four or five or six dial standard resistance box and R_x is the unknown resistance which is to be measured by means of a null method. By adjusting R_3 it can be arranged that no current flows through G.

THE BALANCED CONDITION

Let I be the total current drawn from the d.c. supply E, and let I_1 and I_2 be the currents that will flow through R_1 and R_x, and R_2 and R_3 respectively by adjusting R_3 so that no current flows through G, the pointer is at the centre of the instrument. The p.d. across CD is zero and the potential at C is equal to the potential at D.

p.d. across AC = p.d. across AD

$$I_1 R_1 = I_2 R_2 \qquad \text{... (1)}$$

p.d. across CB = p.d. across BD

$$I_1 R_x = I_2 R_3 \qquad \text{... (2)}$$

Dividing equations (1) and (2)

$$\frac{I_1 R_1}{I_1 R_x} = \frac{I_2 R_2}{I_2 R_3}$$

$$\frac{R_1}{R_x} = \frac{R_2}{R_3} \qquad \text{... (3)}$$

transposing equation (3)

$$R_x = \left(\frac{R_1}{R_2}\right) R_3$$

$\dfrac{R_1}{R_2}$ is called the *arm's ratio*

R_3 is multiplied by R_1 (R_1 is the *multiply*)

R_3 is divided by R_2 (R_2 is the *divide*)

R_1 can be one of three resistors. $R_1 = 10\ \Omega$, $R_1 = 100\ \Omega$, $R_1 = 1{,}000\ \Omega$ R_2 can be one of three resistors $R_2 = 10\ \Omega$, $R_2 = 100\ \Omega$, $R_2 = 1{,}000\ \Omega$ R_3 is four, a five or a six dial variable resistance box.

UNITS	TENS	HUNDREDS	THOUSANDS	TEN THOUSANDS
× 1 Ω	× 10 Ω	× 100 Ω	× 1,000 Ω	× 10,000 Ω

HUNDRED THOUSANDS
× 100,000 Ω

The basic principle of the Wheatstone bridge depends on the balanced equation

$$R_x = \left(\frac{R_1}{R_2}\right) R_3.$$

For a resistance of 1.23 Ω

$$R_x = \left(\frac{10}{1{,}000}\right) 123 = 1.23\ \Omega$$

and for a resistance of 100,000,000 Ω

$$R_x = \left(\frac{1,000}{10}\right) 1,000,000 = 100,000,000 \; \Omega$$

and thus the wheatstone bridge can accurately measure resistance over a wide range.

EXERCISE 8

1. Sketch the basic construction of a moving coil instrument, and explain the principle of operation.

2. A moving-coil milliammeter has a coil resistance of 10 Ω and gives a full scale deflection of current of 15 mA.

    (i)   Calculate the value of the resistor required to enable the instrument to read 0-5 A.

    (ii) Calculate the value of the resistor required to enable the milliammeter to read 0-500 V. (Ans.   (i) 0.03 Ω (ii) 33,323.33 Ω).

3. A moving coil meter movement has a coil resistance of 400 Ω and a resistor of 600 Ω is connected in series with it. The movement requires a current of 2 mA for f.s.d. (full scale deflection). Calculate the shunt resistance required to change the movement into a 0-1 A ammeter.

    (Ans. 2.004 Ω).

4. Explain, with the aid of a suitable diagram the principle of operation of the Wheatstone bridge. Derive from first principles an expression giving, for the balanced condition, the unknown resistance in terms of the known quantities.

5. A 0 to 50 V meter has resistance of 500,000 Ω what additional voltage multiplier resistor is required to enable the meter to indicate 0 to 500V.

    (Ans. 4.5 MΩ).

6. For the Wheatstone bridge shown in Fig. 201.

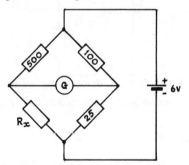

Fig. 201   Wheatstone bridge.

Determine: (i) the value of R_x

                     (ii) the p.d. across the 100 Ω

                     (iii) the current through the 500 Ω resistor.

The bridge is at balance

(Ans. (i)   125 Ω (ii) 4.8 V   (iii) 9.6 mA).

7. Explain with the aid of a circuit diagram what is meant by a potential divider and show how two such potential dividers make up a wheatstone bridge circuit.

8. Compare the Wheatstone bridge with the ammeter and voltmeter methods for measuring resistance.

9. An ammeter of resistance 0.05 Ω has a shunt of resistance 0.015 Ω connected across it. What is the total current flowing when the current through the ammeter is 5 A?

(Ans. 21.7 A).

10. A voltmeter with resistance of 1,000 Ω is used to measure the p.d. between points A and B of the circuit shown in Fig. 202. What is the voltmeter reading and how does this compare with the true value.

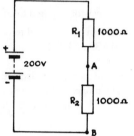

Fig. 202   voltmeter measurement.

(Ans. 66.7 V).

11. An ammeter and a voltmeter are used to check the resistance of a resistor which is marked 1 KΩ. The circuit is connected as in Fig. 203, if the voltmeter has a resistance of 1 KΩ find the resistance value from the calculated instrument readings.

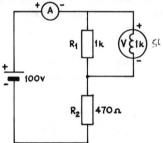

Fig. 203   Ammeter/voltmeter method for measuring resistance.

The resistance of the ammeter may be assumed to be negligible. State the error in the calculated value.

(Ans. 51.55 V, 6.48 V).

12. A moving coil instrument gives full-scale deflection when the current through it is 10 mA and the p.d. across it is 5 V.

(i)   Calculate the resistance of the instrument.

(ii) Find the value of resistor required to enable the instrument to read 38 V at full-scale deflection and illustrate your solution with a diagram.

(iii) Find the value of resistor required to enable the instrument to read 10 A at full scale deflection and illustrate your solution with a diagram.

(Ans. (i) 500 Ω   (ii) 3.3 KΩ   (iii) 0.505 Ω).

13. Construct a simple ohmmeter using an ammeter with a full-scale deflection of 15 mA a variable resistor, and a 15 V battery. Draw the circuit for the arrangement.

Calculate the resistance indicated by the instrument when currents of (i) 1mA (ii) 5 mA and (iii) 10 mA are flowing in the circuits.

(Ans. (i) 14 KΩ   (ii) 2 KΩ   (iii) 0.5 KΩ).

14. Suggest a suitable method of measuring each of the following:

    (i)   The r.m.s value of a 100 Hz sinusoidal voltage of the order of 10 V.

    (ii) The r.m.s. value of a 250 KHz sinusoidal voltage of the order of 100 mV.

    (iii) The peak value of a voltage pulse of rectangular waveform of duration about 20 μs and repeating every 100 μs.

SOLUTION PART II

SOLUTIONS 1

1. The total resistance is calculated from the formula

$$\frac{1}{R_T} = \frac{1}{R_1} + \frac{1}{R_2} + \frac{1}{R_3}$$

$$\frac{1}{R_T} = \frac{1}{1} + \frac{1}{2} + \frac{1}{3} = \frac{6+3+2}{6} = \frac{11}{6}$$

$$R_T = \frac{6}{11}\ \Omega.$$

Fig. 204 shows the circuit.

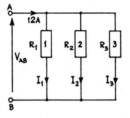

Fig. 204   Parallel resistors.

The p.d. between A and B, $V_{AB} = 12 \times \frac{6}{11} = 6.55$ V. The currents I_1, I_2 and I_3 can be determined $I_1 = V_{AB}/1 = 6.55$ A, $I_2 = V_{AB}/2 = 3.28$ A, $I_3 = V_{AB}/3 = 2.18$ A.

2. The circuit is shown in Fig. 205

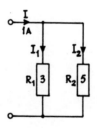

Fig. 205   Current divider.

Using the current divider formulae

$$I_1 = \frac{R_2}{R_1 + R_2}\ I = \frac{5}{8} \times 1 = \frac{5}{8} = 0.025\ \text{A}.$$

$$I_2 = \frac{R_1}{R_1 + R_2}\ I = \frac{3}{8} \times 1 = \frac{3}{8} = 0.375\ \text{A}.$$

3. Fig. 206 shows the circuit

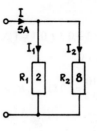

Fig. 206   Current divider.

$$I_2 = \frac{R_1}{R_1 + R_2} \ I = \frac{2}{10} \times 5 = 1 \text{ A} \qquad I_1 = \frac{R_2}{R_1 + R_2} \ I = \frac{8}{10} \times 5 = 4 \text{ A}.$$

4. Fig. 207 shows the circuit.

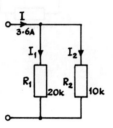

Fig. 207   Current divider.

$$I_1 = \frac{10}{10 + 20} \times 3.6 = \frac{1}{3} \times 3.6 = 1.2 \text{ A}$$

$$I_2 = \frac{20}{10 + 20} \times 3.6 = \frac{2}{3} \times 3.6 = 2.4 \text{ A}.$$

5. $\dfrac{1}{R_T} = \dfrac{1}{30} + \dfrac{1}{40} + \dfrac{1}{50} = \dfrac{20 + 15 + 12}{600} = \dfrac{47}{600}$

$$R_T = \frac{600}{47} = 12.8 \ \Omega.$$

The p.d. across the input = $12.8 \times 2 = 25.6$ V

$$I_1 = \frac{25.6}{30} = 0.853 \text{ A}, \quad I_2 = \frac{25.6}{40} = 0.04 \text{ A}, I_3 = \frac{25.6}{50} = 0.512 \text{ A}.$$

6. $\dfrac{1}{R_T} = \dfrac{1}{3} + \dfrac{1}{7} + \dfrac{1}{R} = \dfrac{1}{1}$

$$\frac{1}{R} = 1 - \frac{1}{3} - \frac{1}{7} = 0.5238$$

$$R = \frac{1}{0.5238} = 1.91 \ \Omega.$$

V_T = p.d. across $R = 10 \times 1.91 = 19.1$ V

$$I_1 = \frac{19.1}{3} = 6.37 \text{ A} \qquad I_2 = \frac{19.1}{7} = 2.73 \text{ A}$$

$$\frac{V}{R_T} = I = 6.37 + 2.73 + 10 = 19.1 \text{ A.}$$

7.   The total current $= I = 0.1 + 0.2 + 0.3 + 0.4 = 1$ A.

The total voltage $= 1 \times 10 = 10$ V

$$R_1 = \frac{10}{0.1} = 100 \ \Omega \qquad R_2 = \frac{10}{0.2} = 50 \ \Omega, \qquad R_3 = \frac{10}{0.3} = 33.3 \ \Omega, \qquad R_4 = \frac{10}{0.4} = 25 \ \Omega.$$

8.   Fig. 208 shows the circuit.

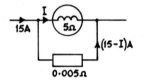

Fig. 208   Shunt.

$5I = (15 - I) \, 0.0005$

$5I + 0.0005 \, I = 0.0075$

$$I = \frac{0.0075}{5.0005} = 1.49985 \times 10^{-3}$$

$I = 1.5$ mA approximately.

9.   Fig. 209 shows the circuit

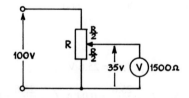

Fig. 209   Loaded potential divider.

$$\frac{\frac{R}{2} \times 1,500}{\frac{R}{2} \times 1,500} \, I = 35 \qquad \ldots (1)$$

$$I \left(\frac{R}{2} + \frac{\frac{R}{2} \times 1,500}{\frac{R}{2} \times 1,500} \right) = 100 \qquad \ldots (2)$$

Dividing equation (2) by equation (1) we have

$$\frac{100}{35} = \frac{R}{2} + \frac{\dfrac{\frac{R}{2} \times 1{,}500}{\frac{R}{2} + 1{,}500}}{\dfrac{\frac{R}{2} \times 1{,}500}{\frac{R}{2} + 1{,}500}}$$

$$= \frac{\dfrac{R}{2} \times 1{,}500}{1{,}500} + 1$$

$$\frac{100}{35} = \frac{R}{2 \times 1500} + 1 + 1$$

$$\frac{100}{35} - 2 = \frac{R}{3{,}000}$$

$$R = 2{,}571.4 \ \Omega \text{ hence } R \approx 2{,}570 \ \Omega.$$

10. $R_1 \dfrac{10{,}000}{R_1 + 10{,}000} I_1 = 80$ ... (1)

$R_2 I_1 = 120$ ... (2)

Dividing (1) by (2)

$$\frac{R_1 \ 10{,}000}{(R_1 + 10{,}000) \ R_2} = \frac{2}{3} \qquad \text{... (3)}$$

$$I_2 = \frac{R_2 \times 10{,}000}{R_2 + 10{,}000} = 100 \qquad \text{... (4)}$$

$$I_2 R_1 = 100 \qquad \text{... (5)}$$

Dividing (4) by (5)

$$\frac{R_2 \ 10{,}000}{R_1 \ (R_2 + 10{,}000)} = 1 \qquad \text{... (6)}$$

From (3) $30{,}000 \ R_1 = 2 R_1 \ R_2 + 20{,}000 \ R_2$ ... (7)

From (6) $10{,}000 \ R_2 = R_1 \ R_2 + 10{,}000 \ R_1$ ... (8)

Substituting $2 \times$ (8) in (7)

$$30{,}000 \ R_1 = 2 \ R_1 R_2 + 2 \ R_1 R_2 + 20{,}000 \ R_1$$

$$10{,}000 \ R_1 = 4 \ R_1 R_2$$

$$R_2 = 2{,}500 \ \Omega$$

and substituting this value in (8)

$$10{,}000 \times 2{,}500 = R_1 \ 2{,}500 + 10{,}000 \ R_1$$

$$10{,}000 \times 2{,}500 = 12{,}500 \ R_1$$

$$R_1 = \frac{10{,}000 \times 2{,}500}{12{,}500} = 2{,}000 \ \Omega.$$

11. $V = \dfrac{5}{30} \times 10 = \dfrac{5}{3} = 1.67$ V.

12. $V = \dfrac{5,000}{30,000} \times 100 = \dfrac{50}{3} = 16.7$ V.

13. Fig. 210 shows the circuit.

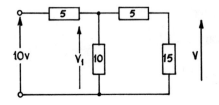

Fig. 210 Ladder network.

$\dfrac{20 \times 10}{30} = \dfrac{20}{3}$

$V_1 = \dfrac{20/3}{5 + 20/3} \times 10 = \dfrac{20/3}{35/3} \times 10 = \dfrac{200}{35}$

$V = \dfrac{15}{20} \times \dfrac{200}{35} = \dfrac{150}{35} = 4.29$ V.

14.

I(A)	6	4.8	4	3.43	3	2.67	2.4	2.18	2
R(Ω)	0	$\frac{1}{4}$	$\frac{1}{2}$	$\frac{3}{4}$	1	$1\frac{1}{4}$	$1\frac{1}{2}$	$1\frac{3}{4}$	2
P(W)	0	5.76	8	8.82	9	8.89	8.64	8.33	8

For $R = \dfrac{1}{4}$ Ω, $I = \dfrac{E}{r + R} = \dfrac{6}{1 + 1/4} = 4.8$ A

$\qquad\qquad P = I^2 R = 4.8^2 \left(\dfrac{1}{4}\right) = 5.76$ W

For $R = \dfrac{3}{4}$ Ω, $I = \dfrac{E}{r + R} = \dfrac{6}{1 + 3/4} = 3.43$ A

$\qquad\qquad P = I^2 R = 3.43^2 \left(\dfrac{3}{4}\right) = 8.82$ W

For $R = 1\dfrac{1}{4}$ Ω, $I = \dfrac{E}{r + R} = \dfrac{6}{1 + 1^1/4} = 2.67$ A

$\qquad\qquad P = I^2 R = 2.67^2 \left(1\dfrac{1}{4}\right) = 8.89$ W

For $R = 1\dfrac{3}{4}$ Ω, $I = \dfrac{E}{r + R} = \dfrac{6}{1 + 1^3/4} = 2.18$ A

$\qquad\qquad P = I^2 R = 2.18^2 \times 1\dfrac{3}{4} = 8.33$ W

Fig. 211 shows the graph of P against R.

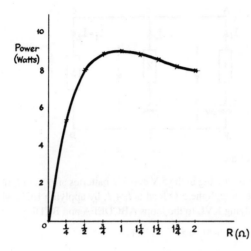

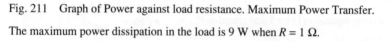

Fig. 211   Graph of Power against load resistance. Maximum Power Transfer.

The maximum power dissipation in the load is 9 W when $R = 1\ \Omega$.

Therefore the maximum power dissipation occurs in the load when the load resistance is equal to the internal resistance.

15. Referring to Fig. 41, the p.d. across the 6 Ω resistor = 6V the current through the 6 Ω resistor = $\dfrac{6}{6} = 1$ A

$I_1 = 1$ A

the total current $= I = I_1 + 0.5 = 1 + 0.5 = 1.5$ A the p.d. across $R = 15 - 6 = 9$ V   therefore $R = \dfrac{9}{1.5} = 6\ \Omega$.

16. (a) $R_{AB} = R_1 + \dfrac{R_2 R_3}{R_2 + R_3} + \dfrac{R_4 R_5 R_6}{R_4 R_5 + R_4 R_6 + R_5 R_6} = 1 + \dfrac{1 \times 1}{1 + 1} + \dfrac{3 \times 3 \times 3}{3 \times 3 + 3 \times 3 + 3 \times 3}$

$= 1 + \dfrac{1}{2} + 1 = 1 + \dfrac{1}{2} + 1 = 2.5\ \Omega$.

(b) (i) $I_1 = \dfrac{10}{R_{AB}} = \dfrac{10}{2.5} = 4$ A

(ii) $I_2 = \dfrac{I_1}{2} = 2$ A

(iii) $I_3 = \dfrac{I_1}{3} = \dfrac{4}{3} = 1.3$ A.

SOLUTION 2

1. Redrawing Fig. 71, we have Fig. 212 showing the currents and p.ds, through and across respectively.

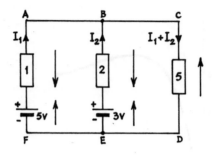

Fig. 212   Kirchhoff's Laws.

Assume that the currents flowing in the 5 V and 3 V batteries are I_1 and I_2 respectively, and hence the load current through the 5 Ω load is $I_1 + I_2$ by applying KCL. Mark the p.ds by arrows as shown. Applying KVL to the loops ABCDEFA and BCDEB, we have

$5 - I_1 - (I_1 + I_2)\, 5 = 0$ or $6I_1 + 5I_2 = 5$ ... (1)

$3 - 2I_2 - (I_1 + I_2)\, 5 = 0$ or $-5I_1 - 7I_2 = -3$ ... (2)

Multiplying equation (1) by 5 and equation (2) by 6, we have

$30I_1 + 25I_2 = 25$ ... (3)

$-30I_1 - 42I_2 = -18$ ... (4)

Adding equations (3) and (4)

$-17I_2 = 7$

$$I_2 = -\frac{7}{17} = -0.41 \text{ A}$$

the negative sign indicates that the assumed current I_2 is flowing in the opposite direction. Substituting $I_2 = -0.41$ A in equation (1)

$6I_1 + 5I_2 = 5$

$6I_1 - 5 \times 0.41 = 5$

$6I_1 = 7.05$

$I_1 = 1.175$ A

the assumed direction of I_1 is correct, therefore I_1 is the discharging current.

$I_1 + I_2 = 1.175 - 0.41 = 0.765$ A.

Therefore $I_1 = 1.175$ A discharging current

$I_2 = -0.41$ A charging current

$I_1 + I_2 = 0.765$ A the load current.

2. Fig. 213 is the redrawn circuit showing the currents and voltages.

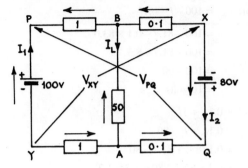

Fig. 213 Kirchhoff's Laws.

(a) Let I_1 and I_2 be the direction of the currents in the batteries and I_L is the load current.

At A, $I_L + I_2 = I_1$, therefore $I_L = I_1 - I_2$.

Applying KVL to the loops PBAYP and BXQAB we have for the loop PBAYP

$$100 = I_1 + (I_1 - I_2)\,50 + I_1$$

$$\text{or } 100 = 52\,I_1 - 50I_2$$

simplifying $50 = 26I_1 - 25I_2$ ... (1)

for the loop BXQAB

$$80 + (I_1 - I_2)\,50 = 0.1I_2 + 0.1I_2$$

$$80 = -50I_1 + 50.2I_2$$

$$\text{or } 40 = -25I_1 + 25.1I_2 \qquad \text{... (2)}$$

Multiplying equation (1) by 25 and equation (2) by 26 we have

$$1{,}250 = 650I_1 - 625I_2 \qquad \text{... (3)}$$

$$1{,}040 = -650I_1 + 652.6I_2 \qquad \text{... (4)}$$

adding (3) and (4)

$$2{,}290 = 27.6I_2$$

$$I_2 = 82.97 \text{ A}$$

substituting this value in (1)

$$50 = 26I_1 - 25 \times 82.97$$

$$26I_1 = 50 + 2074.25$$

$$I_1 = \frac{2124.25}{26} = 81.7 \text{ A}$$

$I_L = I_1 - I_2 = 81.7 - 82.97 = -1.27$ A.

The assumed direction of the load current is incorrect –1.27 A flows in the opposite direction indicated, through the 50 Ω. Therefore $I_1 = 81.7$ A, $I_2 = 83.0$ A, $I_L = -1.27$ A

(b) Let V_{PQ} be the p.d. between P and Q and let V_{XY} be the p.d. between X and Y.

Applying KVL to the loops PQAYP and XQAYX, we have

$$1I_1 + 0.1I_2 + V_{PQ} = 100$$

therefore $V_{PQ} = 100 - I_1 - 0.1I_2$

$$= 100 - 81.7 - 0.1 \times 83.0$$
$$= 100 - 81.7 - 8.3$$
$$= 100 - 90$$
$$= 10 \text{ volts}$$

$$V_{XY} + 80 = 0.1I_2 + 1I_1$$
$$V_{XY} = 0.1I_2 + I_1 - 80$$
$$= 0.1(83) + 81.7 - 80$$
$$= 8.3 + 81.7 - 80 = 8.3 + 1.7$$
$$= 10 \text{ volts.}$$

3. Fig. 73 is now redrawn and it is equivalent to Fig. 214.

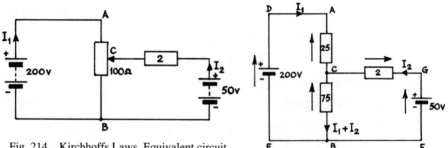

Fig. 214   Kirchhoffs Laws. Equivalent circuit.

Applying KVL to the loops DACBED and CGFBC we have

DACBED $\quad 200 = 25I_1 + 75(I_1 + I_2)$

CGFBC $\quad\quad 50 = 2I_2 + (I_1 + I_2)75$

simplifying $100I_1 + 75I_2 = 200$

dividing each term by 25

$$4I_1 + 3I_2 = 8 \qquad \dots (1)$$
$$75I_1 + 77I_2 = 50 \qquad \dots (2)$$

Multiplying (1) by 75 and (2) by (−4)

$$300I_1 + 225I_2 = 600 \qquad \dots (3)$$
$$-300I_1 - 308I_2 = -200 \quad \dots (4)$$

Adding (3) and (4)

$$-83I_2 = 400$$
$$I_2 = -4.82 \text{ A}$$

substituting this values in (1)

$$4I_1 + 3(-4.82) = 8$$
$$4I_1 = 8 + 3(4.82)$$
$$4I_1 = 2.25$$
$$I_1 = 5.61 \text{ A.}$$

4.   Fig. 215 shows the direction of currents and the p.ds.

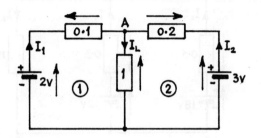

Fig. 215   Kirchhoff's Laws.

Applying KCL at the junction A

$$I_1 + I_2 = I_L$$

Applying KVL to loop ①

$$2 = 0.1I_1 + I_L$$
or $0.1I_1 + I_1 + I_2 = 2$
or $1.1I_1 + I_2 = 2$ ... (1)

Applying KVL to loop ②

$$3 = 0.2I_2 + I_L$$
or $0.2I_2 + I_1 + I_2 = 3$
$I_1 + 1.2I_2 = 3$ ... (2)

Multiplying equation (1) by 1 and (2) by (–1.1)

$$1.1I_1 + I_2 = 2$$
$$-1.1I_1 - 1.2\,(1.1)\,I_2 = 3(-1.1)$$
or $1.1I_1 + I_2 = 2$ ... (3)
$-1.1I_1 - 1.32I_2 = -3.3$ ... (4)

Adding (3) and (4)

$$-0.32I_2 = 1.3$$
$$I_2 = 4.06 \text{ A}$$

substituting $I_2 = 4.06$ in (2)

$$I_1 = 3 - 1.2\,(4.06)$$
$$I_1 = -1.87 \text{ A}$$

I_1 flows in the opposite direction indicated.

5. Fig. 216 shows the circuit with the currents and the voltages.

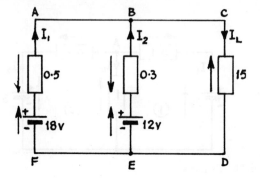

Fig. 216   Kirchhoff's Laws.

At the junction B $I_L = I_1 + I_2$

Applying KVL to the loops ABEFA and BCDEB, we have $18 - 12 = 0.5I_1 - 0.3I_2$

 or $0.5I_1 - 0.3I_2 = 6$ ... (1)

BCDEB $12 = 0.3I_2 + 15 \, (I_1 + I_2)$

 or $15I_1 + 15.3I_2 = 12$ ... (2)

Multiplying equation (1) by –15 and equation (2) by 0.5 we have

   $-7.5I_1 + 4.5I_2 = -90$ ... (3)
   $7.5I_1 + 7.65I_2 = 6$ ... (4)

Adding equations (3) and (4)

   $12.15I_2 = -84$
   $I_2 = -6.91 \text{ A}$

substituting this value in (1)

   $0.5I_1 - 0.3 \, (-6.91) = 6$
   $0.5I_1 = 6 - 2.07$
   $I_1 = \dfrac{3.93}{0.5} = 7.86 \text{ A}$
   $I_1 = 7.86 \text{ A}$

$I_L = I_1 + I_2 = 7.86 - 6.91 = 0.95 \text{ A}.$

Therefore $I_1 = 7.86 \text{ A}, I_2 = -6.91 \text{ A}$ and $I_L = 0.95 \text{ A}.$

6. Referring to Fig. 50 $I = \dfrac{40 + 25 - 10 - 20}{1 + 2 + 3} = 5.83 \text{ A}.$

7. Referring to Fig. 52, $I_1 + I_2 + I_3 = 0$ hence $I_3 = -I_1 - I_2$ Loop ADBEA

   $3 - 2 + 2I_1 - 5I_2 = 0$ or $2I_1 - 5I_2 = -1$ ... (1)

Loop ADBCA

$$-3 + 5I_2 - 10I_3 + 5 = 0 \text{ or } 5I_2 - 10(-I_1 - I_2) = -2$$
$$\text{or } 10I_1 + 15I_2 = -2 \qquad \ldots (2)$$

Multiplying equation (1) by 3, we have

$$6I_1 - 15I_2 = -3 \qquad \ldots (3)$$
$$10I_1 + 15I_2 = -2 \qquad \ldots (2)$$

adding (3) and (2)

$$16I_1 = -5$$
$$I_1 = -\frac{5}{16} = -0.3125 \text{ A.}$$

Substituting this value in (1)

$$2(-0.3125) - 5I_2 = -1$$
$$5I_2 = -0.625 + 1 = 0.375$$
$$I_2 = 0.075 \text{ A}$$
$$I_3 = -(I_1 + I_2) = -(-0.3125 + 0.075) = 0.2375 \text{ A.}$$

The actual currents flow as shown in the diagram of Fig. 217.

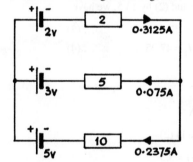

Fig. 217   Actual currents Kirchhoff's Laws.

0.075 A and 0.2375 A are the discharging currents and 0.3125 A is the charging current.

8. Referring to Fig 53 and assuming that the currents flow from the positive terminal of the batteries as shown by I_1 and I_2, then the current through 100 Ω is $I_1 - I_2$ since $I_2 + (I_1 - I_2) = I_1$ at the junction A

Loop ABDEA

$$10 + 100(I_1 - I_2) = 0$$
$$\text{or } 100I_1 - 100I_2 = -10$$
$$\text{or } 100I_2 - 100I_1 = 10$$
$$\text{or } 10I_2 - 10I_1 = 1 \qquad \ldots (1)$$

loop ACBA

$$-30 + 10I_1 - 10 = 0$$
$$40 = 10I_1$$
$$I_1 = 4 \text{ A}$$

substituting in (1)

$$10I_2 - 4\,(10) \ = 1$$
$$10I_2 = 41$$
$$I_2 = 4.1 \text{ A}$$

$$I_1 - I_2 = 4 - 4.1 = -0.1 \text{ A}$$

the current should flow in the opposite direction from that indicated in the diagram.

9. Referring to Fig. 54, loop ADCBA

$$2.5 - 1.5I_1 + (I_1 - I_2)\,15 = 0$$

or $13.5I_1 - 15I_2 = -2.5$

or $15I_2 - 13.5I_1 = 2.5$ ... (1)

Loop ABDA

$$-15\,(I_1 - I_2) + 3.5 - 1.5I_2 = 0$$
$$-15I_1 + 15I_2 + 3.5 - 1.5I_2 = 0$$
$$-15I_1 + 13.5I_2 = -3.5$$

or $15I_1 - 13.5I_2 = 3.5$ ... (2)

Multiplying (1) by 15 and (2) by 13.5, we have

$$-202.5I_1 + 225I_2 = 37.5 \quad\quad ... (3)$$
$$202.5I_1 - 182.25I_2 = 47.25 \quad\quad ... (4)$$

adding (3) and (4)

$$42.75\,I_2 = 84.75$$
$$I_2 = 1.99 \text{ A}$$

Substituting this value in (2)

$$15I_1 - 13.5\,(1.99) = 3.5$$
$$15I_1 = 3.5 + 26.865$$
$$I_1 = 2.02 \text{ A}$$

$$I_1 - I_2 = 2.02 - 1.99 = 0.03 \text{ A}$$

The p.d. across A and B, $V_{AB} = 15 \times 0.03 = 0.45$

10. Referring to Fig. 57, we have

$$I_1 + I_2 + I_3 = 0$$
$$I_3 = -(I_1 + I_2)$$

Applying KVL

Loop EBDAE

$$50 - 20 + 5I_2 - I_1 = 0$$
$$\text{or } I_1 - 5I_2 = 30 \quad\quad ... (1)$$

Loop CBDAC

$$100 - 20 + 5I_2 + (I_1 + I_2)\,0.1 = 0$$
$$0.1I_1 + 5.1I_2 = -80 \qquad \ldots (2)$$

Multiplying (2) by 10

$$I_1 + 51I_2 = -800 \qquad \ldots (3)$$

multiplying (1) by (–1)

$$-I_1 + 5I_2 = -30 \qquad \ldots (4)$$

Adding (3) and (4)

$$56I_2 = -830$$
$$I_2 = -14.8 \text{ A}$$

substituting in (1)

$$I_1 = 30 + 5\,(-14.8)$$
$$= 30 - 74$$
$$= -44 \text{ A}$$

$$I_3 = -(I_1 + I_2) = -(-44 - 14.8) = 58.8 \text{ A}.$$

The currents are $I_1 = -44$ A, $I_2 = -14.8$ A, $I_3 = 58.8$ A the actual currents flow are as shown in Fig. 218

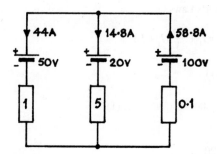

Fig. 218   Kirchhoff's Laws. Actual current.

11. Referring to Fig. 58

$$R_T = 1 + \frac{5 \times 15}{5 + 15} = 1 + \frac{75}{20} = 4.75 \ \Omega$$

$$I = \frac{25}{4.75} = 5.26 \text{ A}$$

using the current divider

$$I_1 = \frac{15}{5+15} \times 5.26 = 3.945 \text{ A}$$

$$I_2 = \frac{5}{20} \times 5.26 = 1.315 \text{ A}$$

or $I_2 = I - I_1 = 5.26 - 3.945 = 1.315$ A

$V_{AB} = 5 \times I_1 = 15 \times I_2 = 5 \times 3.945 = 19.725$ volts

12. Referring to Fig. 59

   The total resistance of the load

   $$= \frac{10 \times 20}{10 + 20} = \frac{20}{3} \ \Omega.$$

   The circuit of Fig. 59 is now equivalent to the circuit of Fig. 219

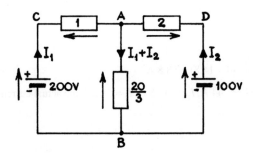

Fig. 219   Kirchhoffs Laws.

Loop CABC $200 = I_1 + \frac{20}{3}(I_1 + I_2)$

$$\frac{23}{3} I_1 + \frac{20}{3} I_2 = 200$$

$$23I_1 + 20I_2 = 600 \qquad \ldots (1)$$

Loop BDAB

$$100 = 2I_2 + \frac{20}{3}(I_1 + I_2)$$

or $\frac{20}{3} I_1 + \frac{26}{3} I_2 = 100$

or $20I_1 + 26I_2 = 300 \qquad \ldots (2)$

Multiplying (1) by 20 and (2) by –23 we have

$$460I_1 + 400I_2 = 12,000 \qquad \ldots (3)$$
$$-460I_1 - 598I_2 = -6,900 \qquad \ldots (4)$$

adding (3) and (4)

$$-198I_2 = 5,100$$
$$I_2 = -25.8 \text{ A}$$

substituting in (1)

$$23I_1 + 20 \,(-25.8) = 600$$
$$23I_1 = 600 + 516$$
$$I_1 = 48.5 \text{ A}$$

13. Referring to Fig. 61, we have

$$I = \frac{25 - 15}{7} = 1.43 \text{ A.}$$
$$V_{AB} = 25 - 2I = 25 - 2 \times 1.43$$
$$V_{AB} = 25 - 2.86 = 22.14 \text{ volts.}$$

14. Referring to Fig. 60, we have. Let 3 I be the supply current, since the circuit is symmetrical, I flows in *AB*, I flows in *AD* and I flows in *AE*, then I/2 flows in *BC* and *BF*, I/2 flows in *DC* and *DH*, I/2 flows in *EF* and *EH* and the total current flows out at *G* as 3 I.

Consider the loop ABCGA

$$V = I \times 1 + \frac{I}{2} \times 1 + I \times 1$$

$$R = \frac{V}{3I} = \frac{1}{3} + \frac{1}{6} + \frac{1}{3} = \frac{2}{3} + \frac{1}{6} + \frac{5}{6} \ \Omega$$

$$R = \frac{5}{6} \ \Omega.$$

15. Fig. 220 shows the circuit required

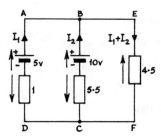

Fig. 220   Kirchhoff's Laws.

Applying KVL to loop ABEFCDA

$$5 = I_1 + (I_1 + I_2)\, 4.5$$
$$5.5I_1 + 4.5I_2 = 5 \qquad \dots (1)$$

Applying KVL to loop BEFCB

$$10 = 5.5I_2 + 4.5\,(I_1 + I_2)$$
$$4.5I_1 + 10I_2 = 10 \qquad \dots (2)$$

Multiplying (1) by 4.5 and (2) by (–5.5), we have

$$24.75I_1 + 20.25I_2 = 22.5 \qquad \ldots (3)$$
$$-24.75I_1 - 55I_2 = -55 \qquad \ldots (4)$$

adding (3) and (4)

$$-34.75I_2 = -32.5$$
$$I_2 = 0.935 \text{ A}$$

substituting this value in (1)

$$5.5I_1 + 4.5 (0.935) = 5$$
$$5.5I_1 = 5 - 4.2075$$
$$I_1 = 0.144 \text{ A}$$
$$I_1 + I_2 = 0.144 + 0.935 = 1.079 \text{ A}.$$

Therefore the branch currents are:

$$I_1 = 0.144 \text{ A}, I_2 = 0.935 \text{ A}, I_1 + I_2 = 1.079 \text{ A}.$$

16. Fig. 221 shows the circuit required

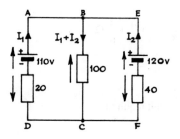

Fig. 221   Kirchhoff's Laws.

KVL ABCDA loop $110 = 20I_1 + 100I_1 + 100I_2$

or $12I_1 + 10I_2 = 11 \qquad \ldots (1)$

Loop BEFCB loop

$$120 = 40I_2 + 100I_1 + 100I_2$$
$$100I_1 + 140I_2 = 120$$

or $10I_1 + 14I_2 = 12$

or $5I_1 + 7I_2 = 6 \qquad \ldots (2)$

Multiplying (1) by 5 and (2) by (−12), we have

$$60I_1 + 50I_2 = 55 \qquad \ldots (3)$$
$$-60I_1 - 84I_2 = -72 \qquad \ldots (4)$$

adding (3) and (4)

$$-34I_2 = -17$$
$$I_2 = \frac{17}{34} = \frac{1}{2}$$
$$I_2 = 0.5 \text{ A}$$

substituting this value in (1)

$$12I_1 + 10 \,(0.5) = 11$$
$$12I_1 = 6$$
$$I_1 = 0.5 \text{ A}$$
$$I_1 + I_2 = 0.5 + 0.5 = 1 \text{ A}.$$

The p.d. across the 100 Ω resistor load $= 1 \times 100 = 100$ V.

17. Fig. 222 shows the circuit.

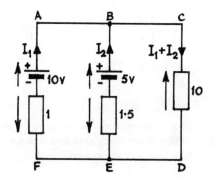

Fig. 222   Kirchhoff's Law.

Applying KVL to the loops ABEFA and BCDEB

Loop ABEFA $10 - 5 + 1.5I_2 - I_1 = 0$

$$\text{or} \quad I_1 - 1.5I_2 = 5 \qquad \qquad \text{... (1)}$$

Loop BCDEB $5 = 1.5I_2 + 10I_1 + 10I_2$

$$\text{or} \quad 10I_1 + 11.5I_2 = 5 \qquad \qquad \text{... (2)}$$

Multiplying (1) by (–10) and (2) by 1, we have

$$-10I_1 + 15I_2 = -50 \qquad \qquad \text{... (3)}$$
$$10I_1 + 11.5I_2 = 5 \qquad \qquad \text{... (4)}$$

adding (3) and (4)

$$26.5I_2 = -45$$

$$I_2 = -\frac{45}{26.5} = -1.7 \text{ A}$$

substituting this value to (1)

$$I_1 - 1.5\,(-1.7) = 5$$
$$I_1 = 5 - 2.55$$
$$I_1 = 3.45 \text{ A}$$
$$I_1 + I_2 = 3.45 - 1.7 = 1.75 \text{ A}$$
$$I_L = 1.75 \text{ A}$$

SOLUTION 3

1. $C_1 = \dfrac{\varepsilon_o \varepsilon_r A_1}{d_1}$ ... (1) $\qquad C_2 = \dfrac{\varepsilon_o \varepsilon_r A_2}{d_2}$ ... (2)

the formulae for the capacitances are given. Dividing (1) by (2)

$$\frac{C_1}{C_2} = \frac{A_1}{A_2}\frac{d_2}{d_1}.$$

If $C_1 = 100$ pF, $\quad C_2 = C_1\dfrac{A_2 d_1}{A_1 d_2} = \dfrac{C_1 d_1 A_1/2}{A_1 2 d_1} = \dfrac{1}{4}C_1$

$$C_2 = \frac{1}{4} 100 \text{ pF} = 25 \text{ pF}$$

$$C_2 = 25 \text{ pF}.$$

2. (a) the capacitance of a capacitor depends directly on the area of the plates, A directly on the dielectric constant of the material placed between the plates ε_r and is inversely proportional to the distance of separations, d.

(b) $C = \dfrac{\varepsilon_o \varepsilon_r A}{d}$

C = capacitance in farads

$\varepsilon_o = 8.85 \times 10^{-12}$ F/m permittivity of free space

ε_r = relative permittivity (no units)

A = area of the plates (m^2)

d = separation (m)

(c) (i) $Q = CV$ Q (coulombs), C (farads), V (volts)

(ii) $W = \dfrac{1}{2} CV^2$ W (joules), C (farads), V (volts)

3. $C = \varepsilon_o A/d$

$5 \times 10^{-12} = \varepsilon_o A/2.5 \times 10^{-3}$ ... (1)

$C = \varepsilon_o A/3.5 \times 10^{-3}$ ... (2)

Dividing (2) by (1)

$$\frac{C}{5 \times 10^{-12}} = \frac{2.5 \times 10^{-3}}{3.5 \times 10^{-3}}$$

$$C = \frac{2.5}{3.5} \times 5 \times 10^{-12} = 3.57 \times 10^{-12} \text{ F}$$

$$C = 3.57 \text{ pF}$$

$C_1 = \dfrac{\varepsilon_o \varepsilon_{r1} A}{d}$ ... (1) $\qquad C_2 = \dfrac{\varepsilon_o \varepsilon_{r2} A}{d}$ ... (2)

A and d are unaffected, dividing (1) by (2) we have $\dfrac{C_1}{C_2} = \dfrac{\varepsilon_{r1}}{\varepsilon_{r2}}$

$$C_2 = C_1 \frac{\varepsilon_{r_2}}{\varepsilon_{r_1}} = 5 \times 10^{-12} \times \frac{5}{1} = 25 \times 10^{-12} \text{ F}$$

$$C_2 = 25 \text{ pF}.$$

Increasing the permittivity, increases the value of the capacitance.

4. $d = 2.5 \text{ mm} = 2.5 \ 10^{-3} \text{ m}$ $A = 100 \text{ cm}^2 = 100 \times 10^{-4} \text{ m}^2$

$C = 100 \text{ pF}$

(i)   electric flux $= Q = CV = 100 \times 10^{-12} \times 15{,}000$

$$= 15 \times 10^{-7} = 1.5 \ \mu C$$

(ii)   electric flux density $= D = Q/A = \dfrac{1.5 \times 10^{-6}}{100 \times 10^{-4}} = 1.5 \times 10^{-4} \text{ C/m}^2$

(iii) $\varepsilon = D/E = \dfrac{1.5 \times 10^{-4}}{15{,}000/2.5 \times 10^{-3}} = \dfrac{1.5 \times 10^{-4}}{6 \times 10^{6}} = 25 \times 10^{-12}$

$$\varepsilon = \varepsilon_r \varepsilon_o = 25 \times 10^{-12} = \varepsilon_r \ 8.85 \times 10^{-12}$$

$$\varepsilon_r = 2.82.$$

5. $W = \dfrac{1}{2} QV = 10^{-6}$

$QV = 2 \times 10^{-6}$

$$V = \dfrac{2 \times 10^{-6}}{100 \times 10^{-6}}$$

$V = 0.02 \text{ volts}$

$$C = Q/V = \dfrac{100 \times 10^{-6}}{0.02}$$

$C = 5{,}000 \ \mu F.$

6.   Fig. 223 shows the circuit of two capacitors in series.

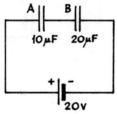

Fig. 223   Two capacitors in series.

(i)   The total charge, Q, is found $Q = CV = 6.67 \times 10^{-6} \times 20$

$$Q = 133.3 \ \mu C \text{ where } C = \dfrac{10 \times 20}{10 + 20} \ \mu F = 6.67 \ \mu F$$

The p.d. across A $= \dfrac{Q}{C_A} = \dfrac{133.3 \times 10^{-6}}{10 \times 10^{-6}} = 13.3 \text{ V}$

The energy stored in $A = \dfrac{1}{2} C_A V_A^2 = \dfrac{1}{2} \times 10 \times 10^{-6} \times 13.33^2$

$$= 888 \ \mu J.$$

(ii) Fig. 224 shows the circuit of two capacitors in parallel.

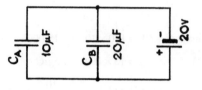

Fig. 224   Two capacitors in parallel.

The p.d. across $C_A = 20$ V.

The energy stored in $A = \dfrac{1}{2} C_A V_A^2$

$$= \dfrac{1}{2} \times 10 \times 10^{-6} \times 20^2$$

$$= 2 \ mJ.$$

7. (i) The equivalent capacitance of the circuit of Fig. 225 is given by

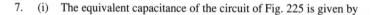

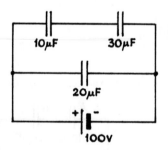

Fig. 225   series-parallel capacitors.

$$C = \left(\dfrac{10 \times 30}{10 + 30} + 20 \right) \ \mu F$$

$$C = 27.5 \ \mu F.$$

(ii) The charge on the 30 μF capacitor is found $Q = \left(\dfrac{10 \times 30}{10 + 30} \right) 100 \times 10^{-6} = 750 \ \mu C$

the p.d. across the 30 μF $= \dfrac{Q}{C} = \dfrac{750 \times 10^{-6}}{30 \times 10^{-6}} = 25$ V.

(iii) The charge on the 10 μF capacitor is the same as that across the 30 μF, 750 μC.

(iv) $W = \dfrac{1}{2} CV^2 = \dfrac{1}{2} \times 20 \times 10^{-6} \times 100^2 = 0.1$ J.

8. (i) Fig. 226 is the circuit required with the appropriate switches.

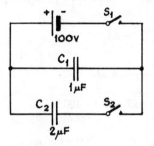

Fig. 226 Charge sharing in capacitors.

S_1 switch is closed and the 1 μF capacitor is charged up to 100 V, S_1 is opened and S_2 is closed what is the p.d. across the two capacitors?

$Q = CV$

$Q = 1 \times 10^{-6} \times 100 = 100$ μC is the charge on C_1.

Let V_o be the new value of the p.d. required

$$V_o = \frac{Q}{C_1 + C_2} = \frac{100 \times 10^{-6}}{3 \times 10^{-6}} = 33.3 \text{ V.}$$

(ii) The energy stored in each capacitor before paralleling

$$W_1 = \frac{1}{2} C_1 V^2 = \frac{1}{2} \times 1 \times 10^{-6} \times 100^2 = 0.005 \text{ J}$$

$$W_2 = \frac{1}{2} C_2 O^2 = O \text{ J}$$

the energy stored in each capacitor after paralleling.

$$W_1 = \frac{1}{2} C_1 V_o^2 = \frac{1}{2} \times 1 \times 10^{-6} \times 33.3^2 = 554 \text{ μJ}$$

$$W_2 = \frac{1}{2} C_2 V_o^2 = \frac{1}{2} \times 2 \times 10^{-6} \ 33.3^2 = 1{,}109 \text{ μJ}$$

Total energy before paralleling 5 mJ.

Total energy after paralleling 1.66 mJ.

A loss of energy $(5 - 1.66)$ mJ = 3.34 mJ due to heat energy in a form of spark.

(iii) Fig. 227 shows the two capacitors in series.

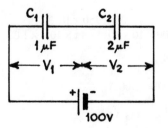

Fig. 227 capacitors in series.

$$Q = CV \quad \left(\frac{1 \times 2}{1 + 2}\right) \mu F \quad \times 100 = \frac{200}{3} \ \mu C = 66.7 \ \mu C$$

$$V_1 = \frac{Q}{C_1} = \frac{66.7 \times 10^{-6}}{1 \times 10^{-6}} = 66.7 \ V$$

$$V_2 = \frac{Q}{C_2} = \frac{66.7 \times 10^{-6}}{2 \times 10^{-6}} = 33.3 \ V$$

therefore $V_1 + V_2 = 66.7 + 33.3 = 100 \ V$.

9. (i) $Q = CV = \dfrac{10 \times 20}{10 + 20} \times 10^{-6} \times 200 = 1.33 \ mC$

(ii) $W = \dfrac{1}{2} \, CV^2 = \dfrac{1}{2} \times \dfrac{10 \times 20}{10 + 20} \times 10^{-6} \times 200^2 = 0.132 \ J$

(iii) $Q = CV$

$V_2 = 60 \ V$ and $V_1 = 140 \ V$

$Q = C_1 V_1 = 10 \times 10^{-6} \times 140 = 1.4 \ mC$

$Q = C'V_2 = 1.4 \ mC$ where $C' = C + C_2$

$C' = \dfrac{1.4}{60} \ mC = 23.3 \ \mu F$

$C = C' - C_2 = 23.3 \ \mu F - 20 \ \mu F = 3.3 \ \mu F$

$C = 3.3 \ \mu F$ is the required value.

10. (i) $C_T = 2C \ // \ C \ // \ 2C = (2C \ // \ 2C) \ // \ C = C \ // \ C = C/2$

(a) $C = 1 \ \mu F, \ C_T = 0.5 \ \mu F$

(b) $C = 10 \ \mu F, \ C_T = 5 \ \mu F$

(ii) $C_T = \dfrac{3C \times C/3}{3C + C/3} = \dfrac{3C^2/3}{10C/3} = \dfrac{3}{10} \, C$

(a) $C = 1 \ \mu F \quad C_T = \dfrac{3}{10} \ \mu F$

(b) $C = 10 \ \mu F \quad C_T = 3 \ \mu F$.

SOLUTIONS 4

1. $\dfrac{IN}{l} = \dfrac{\phi}{A \, \mu_o \mu_r}$

$I = \dfrac{l \, \phi}{NA \, \mu_o \mu_r} = \dfrac{0.5 \times 750 \times 10^{-6}}{300 \times 10 \times 10^{-4} \times 4\pi \times 10^{-7} \times 1{,}000} = 0.9947$

$I = 1 \ A$ approximately.

2. (i) $F = IN = 5 \times 500 = 2{,}500 \ A$

(ii) $H = \dfrac{F}{l} = \dfrac{2{,}500}{0.75} = 3{,}333 \ A/m$

(iii) $B = \phi/A = H\,\mu_o\mu_r$ but a perspex has $\mu_r = 1$

$\qquad B = 3,333 \times 4\pi \times 10^{-7} = 0.00418879$

$\qquad B = 4.19$ mT approximately

(iv) $\phi = BA = 4.19 \times 10^{-3} \times 10 \times 10^{-4} = 4.19\ \mu$ wb.

3. $\quad I = \dfrac{lB}{N\,\mu_o\mu_r} = \dfrac{1 \times 1.5}{1,000 \times 4\pi \times 10^{-7} \times 1,000} = \dfrac{1.5}{4\pi \times 10^{-1}} = 1.19$ A

$\quad S = \dfrac{l_i}{\mu_o\mu_r} + \dfrac{l_a}{\mu_o} = \dfrac{0.999}{4\pi \times 10^{-7} \times 1,000} + \dfrac{0.001}{4\pi \times 10^{-7}} = 795 + 796$

$\qquad = 1,591$ A/Wb

$l_i = 1,000$ mm $- 1$ mm $= 999$ mm $= 0.999$ m

$l_a = 0.001$ m

$S = IN/\phi = IN/BA$ therefore

$B = \dfrac{IN}{AS} = \dfrac{1.19 \times 1,000}{7.5 \times 10^{-4} \times 1,591} = 997$ T rather very high

4. (a) $\quad S = \dfrac{1}{\mu_o\mu_r} = 400,000 = \dfrac{0.628}{4\pi \times 10^{-7}\mu_r}$

$\qquad$ since $D = 20$ cm, $\quad l = 20\ \pi = 0.028$ m

$\qquad \mu_r = \dfrac{0.628}{4p \times 10^{-7} \times 400,000} = 1.25$

(b) $\quad N = 250, \quad I = 1$ A, $\quad l_a = 1$ mm $= 0.001$ m

$\qquad S = \dfrac{l_i}{\mu_o\mu_r} + \dfrac{l_a}{\mu_o} = 400,000 + 400,00 = 800,000$

$\qquad S = 800,000$ A/Wb

$\qquad S = \dfrac{NI}{\phi}$

$\qquad \phi = \dfrac{NI}{S} = \dfrac{250 \times 1}{800,000} = 313\ \mu$ Wb.

5. (a) Fig. 228 shows the B/H curve

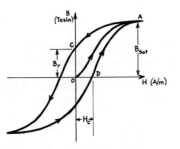

Fig. 228   Hysteresis loop

(i) *OC*

(ii) *OD*

(ii) the initial magnetization curve *OA*

(iv) B_{sat}

(b) $S_i = \dfrac{l_i}{\mu_o \mu_r} = \dfrac{0.5}{4\pi \times 10^{-7} \times 10^3} = 398$ A/Wb

$S = S_i + S_a = 398 + 5 \times 10^5 = 500{,}398$ A/Wb

$S = \dfrac{IN}{\phi} = 500{,}398$ A/Wb

$N = \dfrac{500{,}398 \times 500 \times 10^{-6}}{1.5} = 167$

$N = 167$.

6. Data: $l_i = 0.25$ m, A $= 2 \times 10^{-4}$ m², $N = 1000$ $l_a = 0.001$ m, $\phi = 100 \times 10^{-6}$ Wb

$S = \dfrac{l_i}{\mu_o \mu_r} + \dfrac{l_a}{\mu_o} = \dfrac{0.25}{4\pi \times 10^{-7} \mu_r} + \dfrac{0.001}{4\pi \times 10^{-7}}$

$= \dfrac{0.25}{4\pi \times 10^{-7} \times 177} + \dfrac{0.001}{4\pi \times 10^{-7}} = 1{,}124 + 796 = 1{,}920$

$S = 1{,}920$ A/Wb

where μ_r is found below

$B = 0.5$ T from the magnetisation curve graph corresponding to a value $H = 2{,}250$ A/m

$B = \phi/A = \dfrac{100 \times 10^{-6}}{2 \times 10^{-4}} = 0.5$ T

$\mu_o \mu_r = \dfrac{B}{H} = \dfrac{0.5}{2{,}250}$

$\mu_r = \dfrac{0.5}{2{,}250 \times 4\pi \times 10^{-7}} = 177$

$S = \dfrac{IN}{100 \times 10^{-6}} = \dfrac{I \times 1000}{100 \times 10^{-6}} = 1{,}920$

$I = 192$ μA.

7. (i) $R = \rho \dfrac{l}{A}$

where R is the resistance of copper wire in ohms ρ is the resistivity of the copper wire Ωm, l is the length of the copper wire in m, and A is the cross sectional area in m².

The analogous expression in a magnetic circuit is that of reluctance S.

$S = \left(\dfrac{1}{\mu}\right) \dfrac{l}{A}$

where l is the length of the magnetic path and A is the cross sectional area of the ferromagnetic material where $1/\mu$ is anologous to ρ or μ is analogous to σ, the conductivity.

The units of S are A/Wb, of l is m of A is m^2, and μ is henrys per metre.

(ii) $NI = \phi S = \phi \dfrac{l}{\mu_o \mu_r A}$ where S is the reluctance and $\mu = \mu_o \mu_r$ = absolute permeability.

The analogous expression in an electric circuit is that of ohm's law: e.m.f. = current $\times$ resistance $E = IR$ current is therefore analogous to flux and voltage is analogous to magnetomotive force or ampere-turns in a magnetic circuit.

8. (i) $S = \dfrac{l}{\mu_o \mu_r A} = \dfrac{400 \times 10^{-3}}{4\pi \times 10^{-4} \times 400 \times 10^{-6}} = \dfrac{10^7}{4\pi} = 795{,}775$ A/Wb

(ii) $\phi = \dfrac{NI}{S} = \dfrac{100 \times 5}{795{,}775} = 628\ \mu\text{Wb}$.

9. The reluctance of the air gap is given by

$$S = \dfrac{l}{\mu_o A}$$

where $\mu_r = 1$

$$S = \dfrac{4 \times 10^{-3}}{4\pi \times 10^{-7}\ 400 \times 10^{-6}} = 7{,}957{,}747$$

$$= 7{,}957{,}747\ \text{A/Wb}.$$

The total reluctance

$$S = S_i + S_a$$

S_i = the reluctance of the iron

S_a = the reluctance of the air gap.

$l_i = 400 - 4 = 396$ mm

$S_i = 795{,}775 \times \dfrac{400}{396} = 803{,}813$

$$S = 803{,}813 + 7{,}957{,}747$$

$$S = 8{,}761{,}560\ \text{A/Wb}.$$

The total reluctance.

The effect of introducing the air gap is to increase the reluctance of the circuit. The reluctance of the magnetic circuit with the gap is nearly 11 times larger than the reluctance without the gap and hence the magnetic flux is 11 times smaller

$$\phi = \dfrac{IN}{S} = \dfrac{5 \times 100}{8{,}701{,}560} = 57\ \mu\ \text{Wb}.$$

SOLUTIONS 5

1. $F = BIl$

   where F = force in newtons, B is the flux density in teslas, I is the current in amperes and l = the length of a conductor in metres.

   $$B = \frac{F}{Il} = \frac{0.25}{0.05 \times 0.5} = \frac{0.250}{0.025} = 10 \text{ T}.$$

2. $F = BIl \sin \theta$

   $$= 0.9 \times 0.15 \times 0.4 \times \sin 80° = 0.053 \text{ N}.$$

   The force is now $\left(\frac{1}{5}\right) \times 0.053 = 0.0106 \text{ N}$

   $$0.0106 = 0.9 \times 0.15 \times 0.4 \sin \theta$$

   $$\sin \theta = \frac{0.0106}{0.9 \times 0.15 \times 0.4} = 0.1962963$$

   $$\theta = 11.32°.$$

3. $F = BIl$

   $$I = \frac{F}{Bl} = \frac{100 \times 10^{-3}}{50 \times 10^{-3} \times 10 \times 10^{-2}} = 20 \text{ A}.$$

4. $v = \omega r$ is the relationship connecting angular and linear velocities. The radius of rotation,

   $r = \frac{15}{2} = 7.5 \text{ cm }\; \omega = \frac{2\pi}{60} \times 500 = 52.36 \text{ radians per second}$

   $v = 52.36 \times 0.075 = 393 \text{ m/s}.$

   Fig. 229 shows the rectangular coil

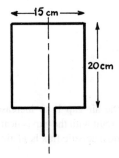

   Fig. 229   Rectangular coil.

   $E = Blv = 1 \times 0.2 \times 3.93 = 0.79 \text{ volts}$

   For 100 turns, $E = 0.79 \times 100 = 79 \text{ volts}.$

5. $E = N \frac{d\phi}{dt}$

   $$N = \frac{E}{d\phi/dt} = \frac{5}{0.5} = 10.$$

6. $E = L \dfrac{di}{dt} = 200 \times 10^{-3} \times 500 = 100$ V.

7. Two coils of self inductance L_A and L_B are placed close to each other as shown in Fig. 230. When the current in L_A changes then the e.m.f. induced across the load resistor, R is $E_2 = M$ di/dt, that is, an induced e.m.f. across R is established where there is a change of current in L_A.

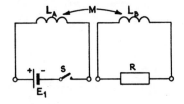

Fig. 230   Mutual inductance for coils close to each other.

When the coils are not placed close to each other, M may be negligible and the e.m.f. induced across R is also negligible. The concept of mutual inductance is well illustrated.

$$E_2 = M \ \dfrac{di}{dt} = 0.2 \times 100 = 20 \text{ V.}$$

8. $\dfrac{E_1}{E_2} = \dfrac{N_1}{N_2} = \dfrac{I_2}{I_1}$ and $R_{in} = \left(\dfrac{N_1}{N_2}\right)^2 R$

$\dfrac{N_2}{N_1} = 10, E_1 = 240$ V and $R = 12$ K Ω.

(i) $E_2 = \dfrac{E_1}{N_1/N_2} = \dfrac{240}{1/10} = 2{,}400$ V note that the transformer is a step up.

$I_2 = \dfrac{E_2}{R} = \dfrac{2{,}400}{12{,}000} = 0.2$ A.

(ii) $I_1 = \dfrac{I_2}{N_1/N_2} = \dfrac{0.2}{1/10} = 2$ A.

(iii) $R_{in} = \left(\dfrac{N_1}{N_2}\right)^2 R = \left(\dfrac{1}{10}\right)^2 12{,}000 = 120 \ \Omega$.

9. (i) $LI = N\phi$ $\qquad\qquad\qquad H = IN/l$

$L = \dfrac{N\phi}{I} = N \ \dfrac{BA}{Hl/N} = N^2 \ \dfrac{BA}{Hl} = N^2 \ \mu_o\mu_r \ A/l$

$L = N^2 \ \mu_o\mu_r \ A/l$

L depends greatly on the number of turns and also depends on μ_r, the relative permeability of the magnetic material.

(ii) $S = \dfrac{l}{\mu_o A}$ reluctance of a non- magnetic material the reluctance is independent of the I

and since S depends on l, μ_o and A which are constants, it is constant.

(iii) $S = \dfrac{l}{\mu_o \mu_r A}$ reluctance of a magnetic material

$\dfrac{B}{H} = \mu_o \mu_r = \dfrac{\phi/A}{IN/l} = \dfrac{\phi l}{AIN}$ therefore

$\mu_r = \dfrac{\phi l}{AIN \mu_o}$ and hence it is inversely proportional to I.

10. See text and Fig. 123

SOLUTIONS 6

1. (i) A non-sinusoidal wave is shown in Fig. 231

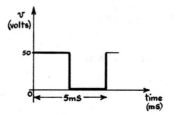

Fig. 231 A non-sinusoidal waveform.

(ii) A sinusoidal current wave is shown in Fig. 232

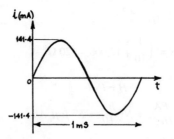

Fig. 232 A sinusoidal waveform.

$I_{max} = 100 \times \sqrt{2}$ mA $= 141.4$ mA

$T = \dfrac{1}{1000}$ 1 ms.

(iii) A sinusoidal voltage wave is shown in Fig. 233

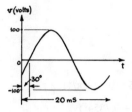

Fig. 233   A lagging sinusoidal voltage wave

$$T = \frac{1}{f} = \frac{1}{.50} = \frac{1000}{50} \text{ ms} = 20 \text{ ms}$$

$30° = \pi/6$ phase angle lagging.

2. $v_1 = 15 \sin \omega t$ is a reference and it is drawn horizontally $AB = 15$ V, $BC = 25$ V. Using the cosine rule to the triangle ABC

$$AC^2 = AB^2 + BC^2 - 2 \times AB \times BC \times \cos 135°$$
$$= 15^2 + 25^2 - 2 \times 15 \times 25 (-0.707)$$
$$= 225 + 625 + 530.25 = 1,380.25$$
$$AC = \sqrt{1,380.25} = 37.15 \text{ V} = \text{is the resultant.}$$

The phase angle of the resultant, ϕ is found by the sine rule

$$\frac{25}{\sin \phi} = \frac{37.15}{\sin 135°}$$

$$\sin \phi = \frac{25 \times \sin 135°}{37.15} = 0.4758457$$

$$\phi = 28.4°$$

$V_R = v_1 + v_2 = 37.15 \sin (\omega t - 28.4°)$

$V_R = 37.15 \sin (\omega t - 28.4°)$

$28.4° = 28.4 \times \dfrac{\pi}{180} = 0.496^c$

therefore $v_R = 37.2 \sin (\omega t - 0.496^c)$.

The phasor diagram of Fig. 234 is drawn to scale by using a ruler, a set square and a protractor. Use 10 V = 1 cm scale. Measure AC and the angle ϕ.

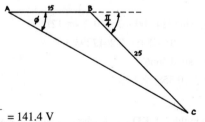

3. (i) $V_{max} = 100\sqrt{2} = 141.4$ V

(ii) $V_{av} = \dfrac{2}{\pi} \times 100 \sqrt{2} = 90$

(iii) $V_{r.m.s.} = \dfrac{100 \sqrt{2}}{\sqrt{2}} = 100$ V

(iv) $\omega = 377$ radians per second

(v) $f = \dfrac{377}{2\pi} = 60$ Hz

(vi) $T = \dfrac{1}{60} = 16.7$ ms

(vii) $v = 141.4 \sin(377 \times 5 \times 10^{-3} + \pi/4)$

$\qquad = 141.4 \sin(1.885^C + 0.7854^C)$

$\qquad = 141.4 \sin 2.67^C$

$\qquad = 64.2$ V

(viii) $\phi = \pi/4$ leading.

$i = \dfrac{v}{100} = \dfrac{10}{100} \sin(500\,\pi\,t - \pi/6)$

$i = 100 \sin(500\,\pi\,t - \pi/6)$ mA

Fig. 235 shows the waveforms of i and v which are in phase.

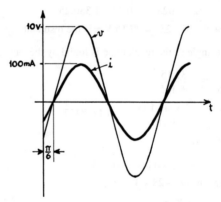

Fig. 235   waveforms in phase.

5. (a) (i) $f = \dfrac{314{,}200}{2\pi} = 50$ KHz

(ii) $I = \dfrac{100 \times 10^{-3}}{\sqrt{2}} = 70.7$ mA

(iii) $i = 100 \times 10^{-3} \sin(314{,}200 \times 2 \times 10^{-6} - \pi/12)$

$\qquad = 100 \times 10^{-3} \sin(0.6284^C - 0.2617293^C)$

$\qquad = 100 \times 10^{-3} \sin 0.3666^C$

$\qquad = 100 \times 10^{-3} \times 0.358$

$\qquad = 35.8$ mA.

(b) Power developed in the 2.2 K$\Omega = I^2 R$ where I is the r.m.s. value of current

$\qquad P = \left(\dfrac{100 \times 10^{-3}}{2}\right)^2 2{,}200 = 11$ W.

(c) Fig. 236 shows that the current lags the voltage by $\pi/12$ or 15°.

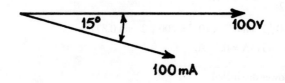

Fig. 236   Phasor diagram.

6. Referring to Fig. 237

E_1 is drawn horizontally representing 10 V = 1 cm, $OA = E_1$, $AB = E_2$, drawn at an angle of 60° leading to the horizontal, $BC = E_3$, drawn at angle of 90° lagging to the horizontal. The resultant voltage is given by $E = 135$ V since $OC = 13.5$ cm and its angle to the horizontal is about 3°.

$E_R = E_M \sin(\omega t + 30°)$ or
$E_R = 135 \sin(\omega t + 0.052^C)$.

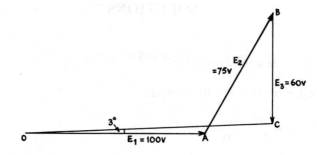

Fig. 237   Phasor diagram.

7. $V = 100 \sqrt{2}$ volts

$f = 50$ Hz

$v = V_{max} \sin \omega t$

$v = 100 \sqrt{2} \ \sqrt{2} \ \sin 2 \pi f t = 200 \sin 2 \pi \times 50 \, t = 100$

$\sin 2 \pi \times 50 \, t = \dfrac{100}{200} = \dfrac{1}{2} = \sin \pi/6$.

$2\pi \times 50 \, t = \pi/6$

$t = \dfrac{1}{600} \times 1000$ ms

$t = 1.67$ ms.

8. $i = 200 \sqrt{2} \sin 877 \, t$

(i) $I = \dfrac{200\sqrt{2} \times 10^{-3}}{\sqrt{2}} = 200$ mA $= 0.2$ A

(ii) $2\pi ft = 877\,t$

$$f = \frac{877}{2\pi} = 139.6 \text{ Hz}$$

(iii) $i = 200\sqrt{2}\,\sin 877 \times 10^{-3} = 200\sqrt{2}\,\sin 0.877^{c}$

$\qquad = 217.43 \text{ mA} = 0.22 \text{ A}.$

9. Average power dissipated

   (i) by calculation

$$P = I^2R = IV = V^2/R$$
$$= 10\sqrt{2} \times 220 = 3{,}110.8 \text{ W}.$$

   (ii) from the waveform of power see graph in text of Fig. 136.

10. $v = 500 \sin \left(2\pi \times \dfrac{1}{10 \times 10^{-3}}\,t + \pi/10 \right)$

$\quad v = 500 \sin (200\,\pi\,t + \pi/10)$ V.

SOLUTIONS 7

1. $X_C = \dfrac{1}{2\pi fC} = \dfrac{1}{2\pi \times 100 \times 100 \times 10^{-6}} = 15.9 \ \Omega.$

2. $X_L = 2\pi fL = 2\pi \times 60 \times 50 \times 10^{-3} = 18.9 \ \Omega.$

3. $X_C = \dfrac{V}{I}$ or $I = \dfrac{V}{X_C} = \dfrac{50}{100} = 0.5$ A.

4. $X_L = \dfrac{V}{I}$ or $V = IX_L = 5 \times 5 = 25$ V.

5. Fig. 238 shows the phase relationships for i, v_L and v_C.

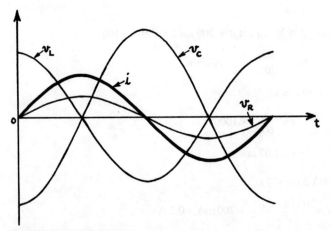

Fig. 238   Waveforms out of phase.

6. Fig. 239 shows the circuit.

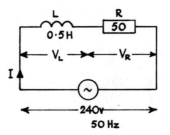

Fig. 239   L-R series circuit.

(i) $Z = \sqrt{R^2 + X_L{}^2} = \sqrt{50^2 + 157.1^2} = 164.9\ \Omega.$

      where $X_L = 2\pi \times 50 \times 0.5 = 157.1\ \Omega$

(ii) $I = \dfrac{240}{164.9} = 1.46\ A$

(iii) $V_L = IX_L = 1.46 \times 157.1 = 229.3\ V$

(iv) $V_R = IR = 1.46 \times 50 = 73\ V$

(v) $V^2 = V_R{}^2 + V_L{}^2 = 73^2 + 229.3^2$

      $V = \sqrt{73^2 + 229.3^2} = 240\ V.$

7. Fig. 240 and Fig. 241 show the d.c. and a.c. circuits of LR in series.

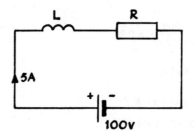

Fig. 240   LR d.c. circuit.

The 100 V is now across R since L is fully magnetised, $R = \dfrac{100}{5} = 20\ \Omega$. The resistance of the coil is 20 Ω.

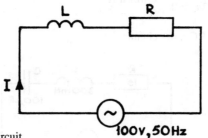

Fig. 241   LR a.c. circuit.

$$I = \frac{V}{Z} = \frac{100}{Z} = 1 \text{ A}$$

$$Z = 100 \ \Omega$$

the impedance of the coil at 50 Hz is 100 Ω.

8. (i) $Z = \sqrt{R^2 + X_L^{\ 2}} = \frac{V}{I} = \frac{200}{10} = 20 \ \Omega$

(ii) $P = IV \cos \phi$

$$\cos \phi = \frac{P}{IV} = \frac{500}{10 \times 200} = 0.25$$

(iii) From the impedance triangle

$$\cos \phi = \frac{R}{Z}$$

$$R = Z \cos \phi = 20 \times 0.25 = 5 \ \Omega.$$

(iv) $Z^2 = R^2 + X_L^{\ 2}$

$X_L^{\ 2} = Z^2 - R^2$

$X_L = \sqrt{Z^2 - R^2} = \sqrt{20^2 - 5^2} = \sqrt{400 - 25} = \sqrt{375} = 19.36$

$2\pi fL = 19.36$

$$L = \frac{19.36}{2\pi \ 50} = 61.6 \text{ mH}.$$

9. (i) $P = IV \cos \phi$

$$\cos \phi = \frac{P}{IV} = \frac{1000}{16 \times 240} = 0.26$$

(ii) $Z = \frac{V}{I} = \frac{240}{16} = 15 \ \Omega$

(iii) From the equation $\cos \phi = R/Z$

$$R = Z \cos \phi = 15 \times 0.26 = 3.9 \ \Omega$$

(iv) $X_L = \sqrt{Z^2 - R^2} = \sqrt{15^2 - 3.9^2} = 14.48 \ \Omega$

(v) $X_L = 2\pi fL$

$$L = \frac{X_L}{2\pi f} = \frac{14.48}{2\pi \times 50} = 46.1 \text{ mH}.$$

10. Fig. 242 shows the LRC series circuit

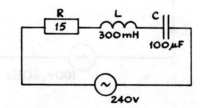

Fig. 242   Series LRC circuit at resonance.

(i) $f_o = \dfrac{1}{2\pi \sqrt{LC}} = \dfrac{1}{2\pi \sqrt{300 \times 10^{-3} \times 100 \times 10^{-6}}}$

$= \dfrac{1000}{2\pi \sqrt{30}} = 29.06$ Hz

(ii) $I_o = \dfrac{240}{15} = 16$ A

(iii) $V_L = I_o \times X_L = 16 \times 2\pi \times 29.06 \times 300 \times 10^{-3} = 876.4$ V

(iv) $V_C = I_o X_C = 16 \times \dfrac{1}{2\pi \times 29.06 \times 100 \times 10^{-6}} = 876.4$ V

(v) $P = I_o^2 R = 16^2 \times 15 = 3,840$ W.

11. The maximum current is the condition of resonance.

Fig. 243 shows the circuit.

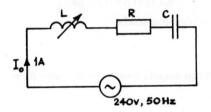

Fig. 243   Resonant series circuit.

$R = \dfrac{V}{I_o} = \dfrac{240}{1} = 240\ \Omega$

$X_L = X_C$ at resonance

$f_o = \dfrac{1}{2\pi \sqrt{LC}}$ squaring up both sides

$LC = \dfrac{1}{4\pi^2 f_o^2} = \dfrac{1}{98696.044} = 1.0132118 \times 10^{-5}$

$V_C = I_o X_C = 1 \times \dfrac{1}{2\pi f_o C} = 500$ hence $C = \dfrac{1}{2\pi \times 50 \times 500} = 6.37\ \mu F$

and $L = \dfrac{1.0132118 \times 10^{-5}}{6.37 \times 10^{-6}} = 1.59$ H.

12. Fig. 244 shows the LRC series circuit

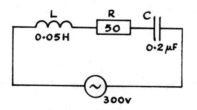

Fig. 244   LRC series resonant circuit.

(i) $f_o = \dfrac{1}{2\pi \sqrt{LC}} = \dfrac{1}{2\pi \sqrt{0.05 \times 0.2 \times 10^{-6}}} = 1{,}591.6$ Hz

(ii) $I_o = \dfrac{300}{50} = 6$ A

(iii) $V_C = I_o X_C = 6 \times \dfrac{1}{2\pi \times 1591.6 \times 0.2 \times 10^{-6}} = 2{,}999.91$

$V_C = 3{,}000$ volts approximately

(iv) $V_C = QV$

$Q = \dfrac{V_C}{V}$ = voltage amplification factor or quality factor

$= \dfrac{3{,}000}{300} = 10$

$Q = 10$.

13. Fig. 245 and Fig. 250 show the circuit and the impedance triangle respectively.

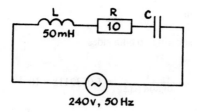

Fig. 245   LRC series circuit.

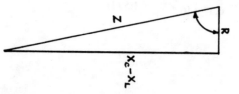

Fig. 246   Impedance triangle.

$$Z = \sqrt{R^2 + (X_C - X_L)^2} = \frac{240}{5} = 48 \ \Omega$$

$$X_L = 2\pi fL = 2\pi \times 50 \times 50 \times 10^{-3} = 157.1 \ \Omega$$

$$(X_C - X_L)^2 = 48^2 - R^2$$

$$X_C - X_L = \sqrt{48^2 - 10^2} = 46.95 \ \Omega$$

$$X_C = 46.95 + 157.1 = 204.05 \ \Omega$$

$$\frac{1}{2\pi \ 50C} = 204.05$$

$$C = \frac{1}{2\pi \times 50 \times 204.05} = 15.6 \ \mu F$$

$$Z = \sqrt{R^2 + (X_L - X_C)^2} = \frac{240}{2} = 120 \ \Omega$$

$$X_L - X_C = \sqrt{120^2 - 10^2} = 119.58261 \ \Omega$$

$$X_L - 119.5826 = X_C$$

$$157.1 - 119.6 = X_C = 37.52 \ \Omega$$

$$X_C = \frac{1}{2\pi \ fC} = 37.52$$

$$C = \frac{1}{2\pi \times 50 \times 37.52} = 84.8 \ \mu F$$

SOLUTIONS 8

1. See in the text constructional details of a moving coil instrument.

2. Fig. 247 shows the shunt circuit

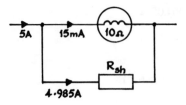

Fig. 247   Shunt circuit Ammeter.

(i)   Let R_{sh} be the resistor required across the 15 mA moving coil milliammeter.

 If the total current is 5 A then the current through R_{sh} is 5 – 0.015 = 4.985 A. The p.d. across the 10 Ω moving coil = 0.015 × 10 = 0.15 V.

 The p.d. across R_{sh} = 4.985 × R_{sh} = 0.15

$$R_{sh} = \frac{0.15}{4.985} = 0.03 \ \Omega.$$

(ii)   Fig. 247a shows the multiplier circuit.

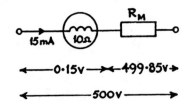

Fig. 247a   Multiplier circuit Voltmeter.

 500 = 0.15 + 0.015 R_M

0.015 R_M = 499.85

$$R_M = \frac{499.85}{0.015} = 33,323.33 \ \Omega.$$

3. Fig. 248 shows the circuit

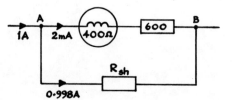

Fig. 248   Shunt circuit.

The p.d. across $AB = 2 \times 10^{-3} \times (400 + 600)$

$$= 2 \text{ V}$$

$$= 0.998 \, R_{sh}$$

$$R_{sh} = \frac{2}{0.998} = 2.004 \, \Omega.$$

4. See text

$$R_x = \left(\frac{R_1}{R_3}\right) R_3 = \text{arm's ratio} \times R_3$$

5. Fig. 249 shows the multiplier circuit

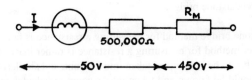

Fig. 249   Multiplier circuit.

$$I = \frac{50}{500,000} = 10^{-4} \text{ A}$$

$$10^{-4} R_M = 450$$

$$R_M = \frac{450}{10^{-4}} = 450 \times 10^4 = 4.5 \text{ M}\Omega.$$

6. (i)   At balance

$$500 \times 25 = 100 \times R_x$$

$$R_x = \frac{500 \times 25}{100} = 125 \, \Omega.$$

(ii)   Current through the 100 Ω resistor $\frac{6}{125} = 48$ mA

p.d. across the 100 $\Omega = 48 \times 10^{-3} \times 100 = 48$ V.

Alternatively

Using the potential divider method p.d. across the 100 $\Omega = \frac{100}{125} \times 6 = 4.8$ V.

(iii)   the current through the 500 $\Omega = \dfrac{6}{500 + R_x} = \dfrac{6}{500 + 125}$

$$= \frac{6}{625} = 9.6 \text{ mA}.$$

7. See text for potential divider at balance, no current flows through G.

$$R_3 R_1 = R_2 R_x$$

$$R_x = \left(\frac{R_1}{R_2}\right) R_3$$

Fig. 250 shows two potential dividers.

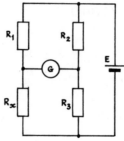

Fig. 250   Wheatstone bridge.

8.   The wheatstone bridge method for measuring a resistance is fairly accurate. The ammeter and voltmeter method for measuring a resistance is rather inaccurate since the p.d. across an ammeter has to be known and taken into account, and similarly the current through the voltmeter should be known and taken into account in calculating accurately the value of resistance.

9.   Applying the current divider method for Fig. 251, we have

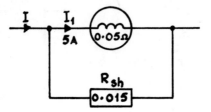

Fig. 251   shunt circuit.

$$I_1 = \frac{0.015}{0.05 + 0.015}\ I$$

$$5 = \frac{0.015}{0.065}\ I$$

$$I = \frac{5 \times 0.065}{0.015} = 21.7\ \text{A}.$$

10.   Fig. 252 shows the potential divider

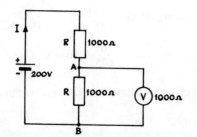

Fig. 252   Potential divider.

If the voltmeter had infinite resistance, the p.d. indicated would have been 100 volts. The p.d. across the voltmeter is found by using the potential divider

$$V = \frac{500}{1500} \times 200 = \frac{200}{3} = 66.7 \text{ V}$$

$$V = 66.7 \text{ V}.$$

11. Again if the voltmeter were ideal, the p.d. across R_1 is found by the potential divider

$$V = \frac{1000}{1470} \times 100 = 68.03 \text{ volts.}$$

The actual p.d. across R_1 is found by lumping the two parallel resistors of 1000 Ω and applying the potential divider

$$V = \frac{500}{970} \times 100 = 51.55 \text{ volts.}$$

The error in the calculated value is $68.03 - 51.55 = 6.48$ volts.

12. (i) From Fig. 253, $R = \dfrac{5}{10 \text{ mA}} = \dfrac{5,000}{10} = 500 \ \Omega$

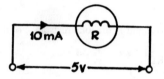

Fig. 253   Moving coil instrument.

(ii) From Fig. 259, $38 = 5 + 10 \times 10^{-3} R_M$

$$R_M \, 10 \times 10^{-3} = 38 - 5$$

$$R_M = \frac{33}{10 \times 10^{-3}} = 3.3 \times 10^3 \ \Omega$$

$$R_M = 3.3 \text{ K}\Omega.$$

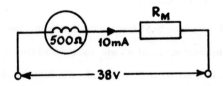

Fig. 254   Multiplier.

(iii) From Fig. 250, $R_{sh} \, 9.9 = 5$

$$R_{sh} = \frac{5}{9.9} = 0.505 \ \Omega.$$

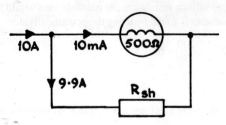

Fig. 255   shunt.

13. Fig. 256 shows the basic circuit of an ohmmeter.

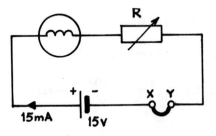

Fig. 256   Ohmmeter.

Short circuit the terminals X and Y, the total resistance of the circuit will be:

$$R = \frac{15}{15 \times 10^{-3}} = 1 \text{ K}\Omega$$

(i)   If $I = 1$ mA, $R = \dfrac{15}{1 \times 10^{-3}} = 15$ KΩ

(ii) If $I = 5$ mA, $R = \dfrac{15}{5 \times 10^{-3}} = 3$ KΩ

(iii) If $I = 10$ mA, $R = \dfrac{15}{10 \times 10^{-3}} = 1.5$ KΩ

In (i) $(15 - 1)$ K$\Omega = 14$ KΩ

In (ii) $(3 - 1)$ K$\Omega = 2$ KΩ

In (iii) $(1.5 - 1)$ K$\Omega = 0.5$ KΩ.

14. (i)   Set the amplifier setting to 5V/cm and the time base setting to 10 ms/cm.

(ii) Set the amplifier setting to 10 mV/cm and the time base setting to 10 μs/cm.

(iii) Set the time base setting to 10 μs/cm and a suitable amplifier setting of 5 V/cm or 1 V/cm.